Principles of
Secure Network
Systems Design

Springer
New York
Berlin
Heidelberg
Barcelona
Hong Kong
London
Milan
Paris
Singapore
Tokyo

Sumit Ghosh

Principles of Secure Network Systems Design

With a Foreword by Harold Lawson

With 91 Figures

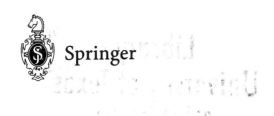

Springer

Sumit Ghosh
Stevens Institute of Technology
Department of Electrical and
   Computer Engineering
Castle Point on Hudson
Burchard Building, Room 211
Hoboken, NJ 07030
USA

Cover illustration: While the nationwide network infrastructure carries traffic from different applications including banking, intelligent transportation systems, military command and control, weapons development labs, patient medical records, and satellite control, the secure transport of each traffic is ensured by the fundamental security framework.

Library of Congress Cataloging-in-Publication Data
Ghosh, Sumit, 1958–
      Principles of secure network systems design / Sumit Ghosh.
         p.   cm.
      Includes bibliographical references and index.
      ISBN 0-387-95213-6 (alk. paper)
      1. Computer networks—Security measures.   2. Computer Security.   I. Title.
   TK5105.59.G64   2002
   005.8—dc21   2001053057

Printed on acid-free paper.

Production managed by Lesley Poliner; manufacturing supervised by Jacqui Ashri.
Photocomposed copy prepared by Texniques, Inc., from LaTeX files supplied by the author.
Printed and bound by Maple-Vail Book Manufacturing Group, York, PA.
Printed in the United States of America.

9 8 7 6 5 4 3 2 1

ISBN 0-387-95213-6   SPIN 10791556

Springer-Verlag   New York Berlin Heidelberg
A member of BertelsmannSpringer Science+Business Media GmbH

To My Loving Family

Foreword

As e-commerce becomes the norm of business transactions and information becomes an essential commodity, it is vital that extensive efforts be made to examine and rectify the problems with the underlying architectures, processes, methods, and tools, as well as organizational structures, that are involved in providing and utilizing services relating to information technology. Such a holistic view of the relevant structures is required in order to identify all of the key aspects that can affect network security.

Unfortunately, today's systems and practices, although they have proved to be useful and become widespread, contain significant unnecessary complexity. This complexity provides many loopholes that make systems and practices vulnerable to malicious attacks by hackers as well as by individual and organized criminals. Further, there are enormous risks due to malfunction of the systems. The holes in the network system cannot simply be plugged up by the use of cryptography and firewalls. While many changes need to be made in operating systems and system software with respect to security, this alone does not solve the problem.

The problems cannot be solved by addressing only a single key aspect of network security. A holistic approach is required. Sumit Ghosh has provided in this book such a holistic view of the area of network security. Thus, it is a most welcome contribution.

The author has a solid background in both the theoretical and practical aspects of network security. As a result, he has been able to see the holistic issues in establishing principles of integrating security into networks as well as to illustrate the deployment of the principles in a concrete case-study example.

The framework presented provides a solid basis for understanding and discussing the issues involved. It is a most appropriate book to use in research and graduate education environments. Further, all of those involved in "architecting" future networks will find this book to be a fundamental contribution in the area. Policymakers as well as practitioners of network security will find a wealth of highly relevant guidance.

cial transactions, online airline reservation systems, worldwide communications, distance learning, and remote data acquisition and control, to name a few. In the near future, NIT systems will encompass innumerable innovative services and products, limited only by our imagination. Fundamentally, in any NIT system, first, the relevant data are identified and acquired from geographically dispersed points; second, the raw data are transformed into usable information through computational intelligence techniques; and third, the information is disseminated to other geographically dispersed regions, as needed. Thus, NIT systems stand on two pillars: computing engines and networking infrastructure. Clearly, the discipline of network security will continue to remain with us for a very long time. In many ways, how well we are able to address the different and sometimes conflicting issues of network security, will determine, to a large extent, the acceptance of the information age by society. A proper balance must be found between the needs of every individual and the collective society, and it must be supported by advances in network-security thinking and technology.

This book is founded on three key objectives. The first is to address the need in the community today for a definition of network security that is fundamental, comprehensive, and applicable to all groups of users including the military, government agencies, industry, and academia. In the past, most networks were isolated, and each utilized its own, limited, vocabulary. In the current environment, however, networks are being connected at an increasingly rapid pace, and multiple groups of users with differing security concerns are being forced to utilize the same network resources concurrently. Previous definitions of network security in the literature have failed to address every aspect of network security, identify the full range of security concerns of different user groups, and permit users to negotiate their respective security issues with each other. Other past efforts offered checklists and specific hardware and software solutions to address a limited set of security concerns. Given the growing scope and complexity of networking today, it is imperative that each user be aware of every aspect of network security and be permitted to exercise the option of focusing on those that are most important or relevant to their needs, budget, and risk assessment. This book presents a framework that addresses, from a fundamental perspective, and in comprehensive manner, every aspect of a network's security.

The second objective is to present an empirical approach to the integration of security into a network's operational and management processes. The goal is to limit the performance degradation that is usually inherited as a result of incorporating security into an operational system. Conceivably, one may exploit the fundamental and unique attributes of a network, if any, to contain the performance degradation of security. To exemplify this approach, a novel security-on-demand system is designed and integrated into the operation of the current asynchronous transfer mode (ATM) networks. This achievement is due principally to the development of the network security framework. This system, in turn, has led to the development of "mixed use" networks, a model for future network design. The expectation is that the reader will analyze the specifics, internalize the approach, and draw upon intuition to design innovative approaches that exploit the unique attributes of future networks, beyond the contemporary IP and ATM networks.

The aim of the third objective is primarily to assess the performance of the network, with security integrated into it, and, secondarily, to verify the second objective. To achieve its goal, the third objective employs accurate behavior modeling of the network's dynamically interacting processes, asynchronous distributed simulation of large-scale representative networks on a loosely coupled network of workstations encapsulating an "almost real-world" testbed, and the analysis of the results through innovative performance metric design. The asynchronous simulation closely mimics reality in that it is subject to events at irregular time intervals. As a result, following proper debugging, the core simulation code may be ported directly, with little modification, to operate actual switches of the network. The author strongly believes that the third objective constitutes a new and practical methodology to gain insights into complex secure network designs.

The key ideas in this book emerged as a result of applying intuition, critical thinking, and reflection on the author's collective research experience as well as those of all the individuals, with whom he has had the privilege of collaborating. The research disciplines include asynchronous distributed algorithms, modeling and simulation of complex systems, hardware systems, performance metric design, networking, and distributed resources allocation. Many of the ideas bear the unmistakable influence of Dr. Jerry Schumacher's (the author's Ph.D. advisee) 19 years of experience in the US military including his tenure as head of the Network Operations Center at the US White House. All of the ideas in the book are supported by current research.

Chapter 1 traces the origin and the historical evolution of security in the fields of communications and networking. The objective is to provide to the reader a perspective on the past that in turn may provide insight into developing a possible path for future evolution. A detailed review of the current literature is also included here. Chapter 1 also presents, from a fundamental perspective, the nature of network security, its principal characteristics, and the key challenges that confront secure network design. It then elaborates a modeling- and simulation-based scientific approach that permits the validation of secure network designs. Chapter 2 focuses on a comprehensive framework for defining network security. Chapter 3 presents an empirical approach to integrating security into the key attributes of a network, utilizing the security framework as the underlying basis. As a case study, it develops a security on demand system for the contemporary ATM network. Chapter 4 addresses the issue of assessing the performance of security integrated into the key attributes of ATM networks through behavior modeling, distributed simulation, and new metric design. It also views the issue of network security as a distributed resource allocation problem and utilizes the dynamic knowledge of the state of the network for higher efficiency in the security-on-demand system. Chapter 5 presents the concept of "mixed use" network that builds on the approaches in Chapters 3 and 4 and promises higher efficiency and significant economies of scale. It serves as a model for future network design. Chapter 6 presents a new and practical approach towards designing secure networks. The approach combines vulnerability analysis, starting from a fundamental and comprehensive understanding of the networking principles, with the design of attacks that aim at exposing the

vulnerabilities, and a modeling- and asynchronous-simulation-based validation of
the synthetic attacks. Armed with insights into the vulnerabilities, the designer
may not only conceive and incorporate network modifications to counteract the
weaknesses, but validate their effectiveness through simulations. Chapter 7 constitutes a refinement of the theme of Chapter 6 in that through creativity and careful
investigation one can uncover design errors in networks even after they have been
meticulously scrutinized by leading international standards organizations. Chapters 6 and 7 underscore a key philosophy of this book. In matters of secure network
design, the perpetrator must never be underestimated. A good designer must bear
in mind that a network design of today may continue to serve society for a very
long time, well beyond its planned life, and therefore one must not spare any effort
in designing it right. One must analyze every conceivable vulnerability, thinking
like a smart perpetrator, eliminate them by careful redesign, and validate the
eventual design through large-scale asynchronous distributed simulation. Finally,
chapter 8 presents a few of the future research issues in network security.

This book has been developed with three types of audiences in mind. First, it is
intended to serve as a primary textbook for a graduate or senior-level undergraduate course titled "Secure Network Design" or "Networked Information Systems
Security" and as a supplementary textbook in a course titled "Network Design."
To facilitate this pedagogical purpose, concepts are developed in this monograph
in a logical, scientific, and canonical fashion. Second, the book targets policymakers and consultants in network security and NIT within industry, military,
law enforcement, and government, and aims to educate them from a fundamental
perspective on the nature, scope, potentials, and limits of network security. The
intensive, short-course style of the book is geared to enable straight-to-the-point,
in-depth self-study. The only prerequisites are basic knowledge of networking, fundamental scientific principles, logical reasoning, common sense, and a serious desire
to learn. Third, it offers to network designers, engineers, practitioners, and managers a systematic yet practical approach to hardening networks against attacks.
As networks increasingly proliferate into the fields of energy, medicine, space, military, banking, transportation, etc., the target audience of this book will increase
steadily.

Sumit Ghosh
Hoboken, New Jersey

Acknowledgments

The author gratefully acknowledges the US Army's support of Dr. Jerry Schumacher's Ph.D. at ASU, members of the networking rating model (NRM) team at the National Security Agency (NSA), and research funding from US Army Research Office, INFOSEC office of NSA, US Air Force Research Labs in Rome (NY), through Motorola Corporation in Scottsdale (AZ), and Sandia National Labs in Albuquerque (NM). The author sincerely thanks his Ph.D. advisees Dr. Jerry Schumacher, currently at US Military Academy at West Point (NY), Dr. Tony Lee, presently at Vitria Technology Inc. in Sunnyvale (CA), Dr. P. Seshasayi, Dr. Ricardo Citro, currently at Intel Corporation in Chandler (AZ), M.S. advisee Aparna Adhav, presently at Cisco Systems in Mountain View (CA), and Ph.D. advisee Prof. Qutaiba Razouqi, presently at Kuwait University, for their research efforts that underlie this book. A number of individuals have contributed greatly to the author's learning, without which this book would not have been possible. These include Dr. Jim Omura of the National Academy of Engineering and formerly of UCLA and Cylink Corporation, Dr. Bud Lawson of Lawson Konsult and inventor of pointers in the discipline of programming languages, Dr. Gottfried Luderer, designer of the first packet switch at AT&T Bell Labs and emeritus professor at ASU, Emily Joyce, Alan Riley, Richard Stevick, and Andy McFarland of NSA, Patricia Edfors formerly of the US Justice Department, Neal Johnson, Dr. Doug Hill, and Ted Simpson of Motorola at Scottsdale (AZ), Tom Tarman and Ed Witzke of Sandia National Laboratories, and Dr. Shukri Wakid, former CIO of NIST and present CIO of National Weather Service/NOAA. For their encouragement, the author expresses sincere thanks to Pete Robinson of US Air Force Research Labs (Rome, NY), Teresa Lunt and O. Sami Saydjari of Darpa, US Army Science Board, Dan Olmos of Boeing Corporation, DISA, Malcolm Airst and Ricard Bailly of the ATM Forum, Ed Woollen of Raytheon Systems Company, Steve Carter of Australian Defense, Bao-Tung Wang of the National Security Bureau of Taiwan, and Elliot Turrini, Assistant US Attorney of the Department of Justice. For understanding and sharing the author's vision, the author expresses his sincere gratitude to members of the administration at Stevens Institute of Technology, including Prof. Stu Tewksbury, Director of the ECE department, Prof. Bernard Gallois, Dean of Engineering, and Dr. Harold Raveche, President. A special note of thanks to Major General Henry J. Schumacher (Ret.) for his encouragement and advice relative to this book. For their continued encouragement and patience with me, Wayne Yuhasz and Wayne Wheeler of Springer deserve my utmost thanks. Last, my sincere thanks to the production staff at Springer.

Sumit Ghosh
Hoboken, New Jersey

Contents

List of Tables

List of Figures

1

Evolution of Network Security and Lessons Learned from History

1.1 Introduction

The business of providing security for the transport of information has been evolving for thousands of years, from the Chinese messenger service through the American Civil War to the two world wars and today's data networks. With the advent of computer networks and, more recently, the increased reliance on networks and associated resources used by the military, government, industry, and academia, the pace of the evolution has greatly accelerated.

The increasing pressure to merge the military and civilian networks into a common network infrastructure has resulted in a dramatic shift in the manner in which security is defined and implemented in a network. Traditionally, each type of user—military, government agency, and civilian company—generated its own requirements and defined its own boundaries of network security. The military actually executed a certification process that upon completion was considered to render the network secure for an extended period of time.

Budget constraints coupled with national security concerns have profoundly influenced today's perspective on network security. Increasingly, the military and industrial sectors are sharing the same network resources to transport information. The classic definitions of network security, used by each sector independently, are clearly inappropriate for the shared resource usage. A new, comprehensive definition of network security is necessitated to satisfy the differing requirements of every user. The definition must enable every sector—military, government, industry, and academia—to discuss and validate security concerns through a unified framework. As a result, not only can one meaningfully assess whether to use an existing network to transport one's own traffic, the framework also allows one to view in a uniform manner the net effect of integrating two or more independent networks.

To exploit the framework for real-world networks, the author proposes a twofold approach: (a) focus on the inherent attributes of a network to take advantage of its unique characteristics, if any, from the perspective of incorporating security into the network, and (b) integrate security into the operation of the network to reduce the impact on performance. The choice of ATM networks for the purpose of illustration underlies their increasing popularity as the network of the future. For

ATM networks the successful integration of (a) and (b) has resulted in a security on demand approach, which enables a user to acquire network resources for the transport of traffic, dynamically and efficiently.

Information networks, the author submits, are inevitably headed toward the notion of "mixed-use" networks. "Mixed-use" is a new concept that is introduced in this book to describe how different users may combine their information network requirements into a single common network infrastructure. However, before a "mixed-use" network may be successfully realized, a general mechanism must be developed that provides for the network security mechanisms of every user type.

Thus, the key advantages of the security framework and the "mixed-use" concept include:

1. Superior resource allocation through distributed resources allocation algorithms

2. Effective cost sharing of security resources, during both development and use

Traditional textbooks on network security focus on specific security hardware and software devices. In contrast, the approach adopted in this book is unique in that it provides a fundamental basis for defining and bounding network security accompanied by a practical methodology for implementing security on demand. Furthermore, this book presents behavior modeling, asynchronous distributed simulation of a large-scale representative network on a loosely coupled parallel processing testbed, and extensive performance analysis as a scientific approach to validating and evaluating the proposed security mechanisms.

1.2 History of Security and Its Influence on Information Networks

Broadly speaking, for any given system, security is determined by its boundaries, which includes people, surrounding environment, time, budget, and perceived threat. In ancient times, the Chinese used specialized messengers to transport messages. These individuals were selected for their speed in memorization, translation, and running. The information was protected either through memorization or hidden on the body when it had to be written down. The ancient Chinese relied less on written messages than on the memorization of information, which was translated by messenger at the destination. When messages had to be written down, they were hidden in a wax ball and either swallowed or hidden on the messenger's body. Today, information security involves less painful methods, although some may be cumbersome, yielding results not too different from those of the ancient messengers. With the progress of time, unique systems and rules for information protection evolved in different societies and especially military systems.

Clausewitz's [1] rules of war and secrecy were created at a time when the world's countries were isolated and very distinct. Today, the world is highly interdependent, and the political boundaries are becoming increasingly blurred. This interdependence coupled with an increasing reliance on computer networks to control critical services that impact our daily lives has greatly elevated the importance of security for these networks. Critical functions of the government, military, and industry rely on networks that are increasingly interconnected with the civilian networks. While this convergence has already changed the way in which security is viewed today, this is a relatively recent occurrence, and the full scope of the problem is yet to be understood fully. The greatest challenge stems from the fact that network security is the culmination of the security issues in many different fields, including computers and communications.

In his book *The Codebreakers* [2], the author traces the origin of modern information security to ancient Egypt, 4,000 years ago, when the first scribe created the new hieroglyphic symbols to be used as a form of shorthand. According to [2], the hieroglyphics are "a deliberate transformation of the writing. It is the oldest text known to do so." Cryptology evolved from these early days, not at a steady pace but in spurts, depending on the civilization and the technology available at the time, but with information security being the persistent driving force.

For a very long time cryptology has remained the central focus of information security. Kahn [2] defines cryptology as the science of rendering signals secure and extracting information from them, and he includes the means to deprive the enemy of information that may be obtained by studying the traffic patterns of radio messages and radar emissions. The emphasis in the definition on the military and its implementation and use of information security is clear.

The scope of cryptology is vast and includes mathematical and linguistic riddles, steganography, polyalphabeticity, monoalphabetic substitution, codebooks, invisible ink, black chambers, variable-key ciphers, wheel ciphers, word transposition, and mechanical and computer devices. An excellent source of information on the history of these techniques is the book *The Codebreakers*.

The history of cryptology is not without its share of failures, the principal causes being either the primitive nature of the techniques or the difficulty in using them. The ancient Chinese are known to have possessed a form of encipherment, and it appears that they rarely used it because it was "neither practical nor effective" [2]. The US Army's experience with a network security system in the early 1990s provides a superb example of how excessive difficulty in using a secure system may lead to its not being used. The system in question involved 10 different login screens, and the user was required to go to a separate room, sometimes located in a different building, and sit under a Plexiglas "cone of silence" before the first screen would appear. The user would then have to pass successively through nine different login screens before gaining access to the system to send and receive information. While the system offered high security, it was cumbersome and slow, and most users bypassed it, resulting in the security being compromised.

The American Civil War witnessed the use of cipher disks for polyalphabetic substitution, originally proposed by the Frenchman Vigenere in 1566 [2]. While the choice appeared initially to be sound, its implementation was a failure. First, the system was not integrated into normal operations and its use was cumbersome. Second, transmission errors caused garbled messages, which led to delays in the transport of critical information. Third, the cipher proved easily vulnerable to intuitive techniques [2].

For an excellent source on the history of cryptography in the US, the reader is referred to the book *The American Black Chamber*, by Herbert O. Yardley, published in 1931. In this book Yardley discloses establishing a code-breaking team during World War I that was capable of reading the coded messages of other nations including Germany and Japan, and that other nations were reading the weak US codes with equal ease. In World War II, two decades later, the situation repeated itself exactly. In August 1945, William P. Bundy led a team that interviewed the German codebreakers. The team learned that the Germans could read the codes of 34 nations including the United States, British Empire, France, Switzerland, China, and Japan [2]. The principal reason the world did not come to know of this fact at the time was that Germany lost the war.

The first known mechanical device to print ciphers was described in detail in 1888 by the Marquis de Viaris [2]. These machines are the predecessors of the modern crypto devices and keys that are in use in information networks even today. The development of different cryptographic techniques and machines led to their widespread use by the military establishments of the world. The United States utilized the unique American Indian languages and their native speakers as human coding machines over the radio. The Navaho codetalkers in the US Marine Corps during World War II grew from 30 at the start of the war to an impressive 420 by the end of the war [2].

During and immediately after World War II, encryption was used extensively to protect the communications lines over which messages were transmitted. Later, when the military built data networks, the practice of encrypting data was carried over to these new networks, constituting the backbone of computer network security in the military. The military started out with the idea of securing each individual computer and later expanded the concept to securing a network of computers and devices. However, it is not the only organization that requires and has implemented some form of security. Network security has evolved over the years, and other departments of government have also embraced the idea of developing secure networks.

The computer-driven integration of the fields of (i) communications and (ii) automation and control are primarily responsible for the proliferation of today's networks. Computer networks have grown from a simple time-sharing system—a number of terminals connected to a central computer—to large, complex environments that provide the infrastructure to many critical and economically valuable components of the economy. Many of the large-scale real-world systems in the government, military, and civilian sectors consist of a number of geographically dispersed hardware and software entities that are interconnected through a network that facilitates the exchange of both data and control traffic. Examples include the US Treasury network, FBI network, Federal Reserve banking network, national power grid, and commercial networks for banks, financial institutions, and credit card transactions.

Today, there is an increased reliance on computer networks, not all of which may be widely known to the general public. In fact, most US residents do not realize that they rely on hundreds of computer networks during the normal course of the day and that the proper functioning of these networks is critical to our well-being and survival. As a result, the risk to the economy, infrastructure, and well-being of the population has not necessarily been widely reported and has only recently been in the spotlight [3]. The report [3] underscores the vulnerabilities of these networks and addresses the issue of security from all perspectives: military, industry and government. Such complex systems, however, are often vulnerable to failures, intrusion, and other catastrophes. Backhouse and Dhillon [4] estimate the yearly damages to the vulnerable finance and banking sectors in the US at $2 billion. With the growing use and ubiquitous reliance on such computer networks, increasing emphasis is being placed on security. Both industry and government are engaged in developing new ways to ensure that the networks are more reliable, survivable, and secure. This increased reliance on networks has caused industry to pay more attention to security requirements.

1.3 Lessons Learned from History

One of the greatest insights into the complex nature of network security is revealed in modern history by the meteoric rise to fame of the German U-boat fleet followed by its spectacular and rapid demise [5][6]. At the beginning of World War II, Admiral Doenitz, architect of the German U-boat fleet, instituted a command and control (C^2) strategy with two key elements. The first was the Enigma encryption system, consisting of the typewriter look-alike Enigma machines and the code book that was used to encrypt and decrypt every radio communication between each U-boat and Doenitz's central headquarters (HQ). The German military scientists believed that Enigma was unbreakable, and this was indeed true at the beginning of the war. One would think that given this invincibility and the fact that directional range finders (DRFs) at the time were unreliable and inaccurate, Doenitz's strategists could rely on Enigma alone to guarantee that the enemy would never learn the exact whereabouts of any U-boat. However, Doenitz's strategists added a second element that is explained as follows. At the time of leaving the base, every individual U-boat commander was assigned by the HQ a mission, namely to destroy allied shipping; a specific territory, say a few hundred square miles of the Atlantic ocean over which to carry out its mission; and a mission duration ranging anywhere from 3 to 6 months. During this interval, there would be no required periodic communication between the U-boat and HQ except when the HQ needed to modify a U-boat's mission or a U-boat required specific instructions from the HQ. Thus, while the communications between the HQ and the U-boats were asynchronous, i.e., irregular in time and unknown a priori, the duration of any communication was short, making it extremely difficult for the allies to locate any U-boat, even as DRFs were continually being improved. The combination of the two elements constituted a brilliant C^2 strategy. It provided true stealth to the U-boats in that no one, not even the HQ, knew of any U-boat's exact location at any time. Doenitz's C^2 strategy underscores a key lesson in that the nature of security is highly complex and requires a holistic approach to guarantee success. Though indispensable, encryption, by itself, may not always be adequate. As World War II progressed, the German high command became increasingly overconfident and started to micromanage the U-boats for even greater efficiency. Every U-boat was required to radio in its present position to the HQ at regular intervals, severely crippling the second element of the strategy. The allies detected this regularity in U-boat to HQ communication and quickly learned to deploy their DRFs at the exact same interval. By this time, the allies had also broken the Enigma code and were able to decipher German communications in real time. Using highly improved DRFs and cross checking the triangulation results against the position reported by an individual U-boat to the HQ, after deciphering the communication, allied destroyers were literally on top of a U-boat minutes after it had started its radio conversation with the HQ. Even if the allies had not successfully broken Enigma, the outcome would have been no different. U-boat losses started to mount, and Doenitz suspected that the allies had successfully broken the Enigma code. He gathered a group of top military scientists. For every destroyed U-boat he provided the scientists with post mortem data on the timings of the U-boat to HQ communications, the message contents, and the time at which it was attacked by the allies and asked them to assess his suspicion through careful analysis. The scientists were so confident of Enigma that they categorically rejected the idea that the code had been compromised. Soon, the U-boat fleet suffered devastating destruction, losing as many as 44 U-boats in a single month, May 1943, and their threat was completely eliminated. Though it is debatable, had the German military scientists been more open-minded about Enigma's vulnerability and advised the HQ to return autonomy to the individual U-boat commanders, it is probable that the allies, despite having broken the Enigma code, would have faced considerable difficulty in locating and destroying the U-boat fleet. In the year 2000, for reasons of US national security and survivability, the only individuals who know the precise location of any given US Trident submarine while it is performing a mission are all on board [7]. Although the navy commander who had originally assigned the few hundred square miles of territory of one of the world's oceans to the submarine commander may have some idea, even he or she does not know the submarine's precise location at any time.

Analysis of this important historic event reveals three key lessons that have influenced the thinking in the remainder of this book. First, no organization should blindly believe that its security is infallible, that it will remain at the cutting edge of communications indefinitely, and that it will maintain its role forever as the sole provider of security. Human creativity is incredible and unpredictable. The second lesson is that what one does with the information that one has acquired from breaking another organization's codes is far more important than the volume of information acquired. In World War II, for example, the US government's failure to act on information received prior to December 7, 1941, had to be redeemed later in the war at the battle of Midway and the Normandy invasion. Ironically, without discipline and proper focus, voluminous amounts of deciphered information may easily overload and overwhelm one's own computational and informational infrastructure. The third lesson is a more serious one, and it relates to the human issue. No amount of technical sophistication in security can ever hope to anticipate and rectify the errors made by individual persons.

Observation of the growing proliferation of information in virtually every aspect of society, coupled with the realization that computational engines and networks constitute the fundamental underlying infrastructures, implies the need for governments to assume a new vision relative to network security. In this vision, every government can and should improve its security in a cost-effective manner, preferably in conjunction with the nation's industry. Government and industry utilize different forms of security, and both of these forms must be improved together. While it may be prohibitively expensive to improve them separately, to keep their developments totally separate may be even more costly to the nation in the long term. The solution, therefore, consists in the following steps:

- Integrate regional and national networks

- Integrate civilian, government, military, and industrial sector networks

- Assign an appropriate level of security to every message generated within every network in the country

- Incorporate better protection for security resources from human failings

1.4 Growing Interest in Network Security

During World War II, the military started encrypting message traffic using mechanical devices to encode and decode. With the advent of computers and communication between computers, the military started to encrypt the information sent over its communications links. The well-researched discipline of cryptography relates to the privacy attribute of a secure network. Network security is multidimensional. While improving the privacy of a network will clearly enhance its security, to ensure comprehensive network security, every attribute must be addressed thoroughly.

Increasingly, the thrust of security is shifting from protecting the data on the computer to securing the information derived from the data and transported through the network. Today, large sectors of the economy rely on computer networks and are susceptible to attacks. The power grid of the United States is vulnerable to mischief. A terrorist attack on the computer network that controls it can bring about catastrophic failure with serious consequences. The nation's financial transaction network; telephone network; credit card transaction network; Department of Defense (DoD) supply, personnel, and pay systems; airline reservation system; the proposed medical network; and the proposed intelligent transportation system (ITS) network are all vulnerable to attacks. The need to protect these networks is real, increasing, and pervasive. To achieve this goal in a systematic manner, against different threats, first the attributes of a secure network must be defined. One must bound what one aims to protect before one can analyze the effectiveness of the protection. The objective, in turn, requires one to first define a language in which the security concerns may be expressed. A key difficulty has been the use of different languages by the military, government, and industry to specify their respective security requirements. In the military, security is organized into COMmunications SECurity (COMSEC) and INFOrmation SECurity (INFOSEC), the definitions of which are included in the Orange and Red books, respectively. The federal government shares the basic principles of the military but utilizes a different language. Industry employs a very different language. The lack of a common language was recognized by the NSA, the nation's proponent for computer and network security, at the first Network Rating Model (NRM) conference in Williamsburg, March 1996. The requirement for developing an NRM has become a key priority.

Recently, commercial industry has become very interested in security of networks. For today, a favorable cost benefit can be associated with security. The Internet's growing popularity and potential for commerce has increased the amount of money and effort devoted to produce and enforce security for privacy and nonrepudiation attributes. Corporations such as General Electric that have lost money due to intrusion can now justify increased attention and spending on network security. To assign a dollar value to security, industry can estimate either the worth of the information compromised or the production time lost while the network is brought down, relative to the money that could have been expended to defeat the attack. For the federal government, network penetration can translate directly into human lives jeopardized, millions of dollars of research investment lost, or costly war plans rendered useless.

Given the geographically distributed characteristic of networks and the increasingly dynamic and asynchronous nature of the interactions between the users and the network, there is an increasing demand for security on demand. Under this approach, one may specify a unique set of security requirements for one's traffic prior to transporting it. Figure 1.1 presents a scenario where, while determining how to combine network A and network B, one observes that none of the four definitions of network security in user group A network overlaps or correlates with those in user group B. The scenario is an accurate reflection of today's reality and is a key obstacle to the integration of military networks and the public network infrastructure.

Both the federal government and industry agree on a need for network security. The federal government has cases of lives being jeopardized and millions of dollars in research or war plans lost due to computer network penetration. Industry can put a dollar value on security by costing out what the information lost is worth or what production time was lost if a network is brought down and comparing how much should be spent on security. There is no debate on whether to protect networks. At issue are the degree of protection that must be provided, the manner in which the protection is evaluated, and the cost and performance tradeoffs of implementing security in a network. One such debate took place at the NSA's NRM conference in 1996 with members from industry, government, military, and academia as participants. The subsequent authors' working group meetings led to

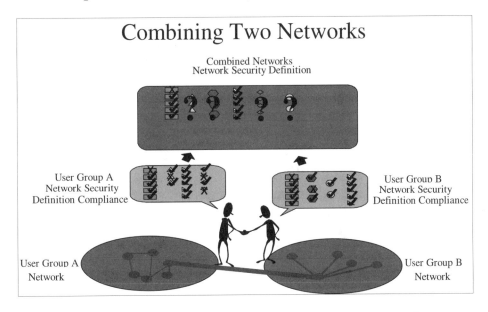

Figure 1.1 Integrating two networks from the security perspective.

the adoption of this book's fundamental framework for network security, shown
in Figure 1.2, as the underlying rationale for the NRM.

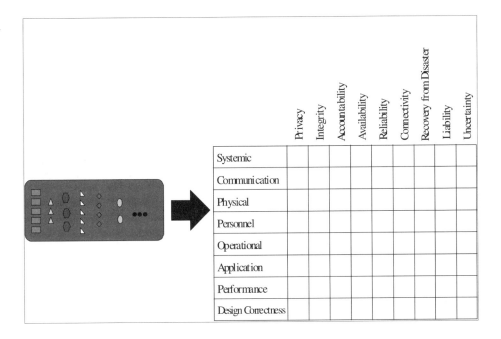

Figure 1.2 Fundamental framework for network security.

Existing and planned networks are subject to threats. Unlike the situation in the past, where the principal threats were directed from a few enemy countries, today the sources of threats are more ubiquitous and the threats themselves more subtle, ingenious, and dangerous. To counter the threats, the security administrator of a network must first examine the perceived threats, assess their potential impact on the key attributes of network security, and then determine the allocation of necessary resources, utilizing the relative importance of the attributes.

1.5 Origin and Nature of Security in Networks

To understand the principles that underlie the security of any network, one must first understand the primary causes that trigger the need for security in networks. Networks are characterized by three fundamental properties: "sharing," "physical distance," and buffers [8]. Sharing underscores the notion that networks are necessarily shared between multiple users, by choice and for reasons of efficiency. This clearly implies the vulnerability of any user's traffic to other users in the network. The notion of physical distance reflects the fact that users are necessarily geographically distributed: No two users reside at the exact same location at the same time. Depending on whether a network spans a VLSI chip or interplanetary space, the measure of physical distance may range anywhere from millimeters to hundreds of thousands of kilometers. Given that a sender and a target receiver are located at a finite distance from each other, a transmitted message cannot remain under the sole control of the sender or receiver at all times during the transmission, and it is therefore clearly vulnerable to internal and external forces that may have access to the intervening portion of the network.

Buffers constitute fast storage devices in switching nodes. In any network, packets require switching along the correct route and up to their final destination, and switching nodes provide the necessary computational intelligence. Upon intercepting a packet, a node processes it and then directs it to an appropriate output port. Clearly, to process it, the underlying computing engine must first store the packet in the buffers, examine it, and then execute its decision relative to the packet's onward transport. If the packet cannot be stored for an appropriate duration, even if brief, it cannot be processed correctly by a computing engine. The need to store it constitutes both a physical and mathematical requirement and is also a direct cause of the packet being vulnerable to any entity that may have forcibly gained control of the node. Once stored, exact duplicates of the packet may be created and then subject to different kinds of mischief. For example, consider "secure" credit card transactions transported over the current Internet where the credit card numbers and other sensitive information are encrypted through state-of-the-art encryption algorithms. As per the definition of IP, the packets may be routed through intermediate routing nodes, not known a priori to the sending node. Assume that at any of these nodes, a determined perpetrator quietly copies the packets from the buffer onto a storage device for subsequent analysis. Although unable to decode the credit card numbers at the present, with gradual improvements in decrypting techniques and computing power, the perpetrator is likely to successfully decode the information in the near future, ranging from a few months to possibly a few years. If and when that occurs, the perpetrator may end up acquiring a very large set of compromised credit card numbers and launch a coordinated attack on every one of the cards, most of which might still be in circulation, thereby causing unprecedented harm. Moreover, attempts by the issuing banks to replace the compromised credit card numbers with new ones may be difficult, given the large number of compromised card numbers and that although the 16-digit card number implies numerable combinations, the set is finite.

The fundamental nature of security is that it is born out of conflicting requirements. First, as observed earlier, every user's packets are vulnerable to other users within the network as well as external forces that may acquire physical access to either a subset or the entire network. At the same time, the complete elimination of the vulnerabilities is precluded by the very definition of networks. Second, on the one hand, a sender transmitting a message M_1 would prefer that M_1 be rendered unreadable or destroyed should it fall into the hands of users other than the target receiver. At the same time, the sender would like to see the purpose underlying the transmittal of M_1 fulfilled. Secure network design must therefore focus on reducing the vulnerability, as well as possible, subject to cost, needs, and other factors.

1.6 Principal Characteristics of Network Security

- 1. Limiting access to a sender-launched packet: Although a packet P_1 launched by a sender S_1 is intended, in general, for a single target user R_1 or a select list of recipient users $R_1, R_2, \ldots R_n$, consider, without any loss in generality, that S_1 and a single R_i are associated with every packet. A secure network must, therefore, prevent P_1 from being accessed by any of the other receivers, as much as possible. Clearly, the use of broadcast oriented techniques as in the Ethernet where any receiver on the network can potentially access P_1 does not constitute a logical decision.

- 2. Encoding packet content to reduce risk of exposure: Recognizing that the network may not succeed with certainty in preventing P_1 from being intercepted by other receivers in the network, logic dictates that the content of P_1 be encoded with an encryption algorithm that is known only to the sender and the designated receiver. Thus, even if P_1 were to be accessed by an unauthorized receiver, it would be denied immediate access to the information contained in P_1.

- 3. Sender notification in the event of nondelivery of a packet: In the event that a packet is not delivered to the target destination within a reasonable time interval, presumably lost or intercepted in transit, the sender should be notified, where possible, so it may adopt appropriate measures to reduce the impact of the exposure. Given that it is aware of the content of the message and its desired effect, the sender is the most logical candidate to execute effective countermeasures. An example from history underscores the importance of this characteristic. Following the assassination of Archduke Franz Ferdinand, heir to the Austro-Hungarian empire, the emperor sent a letter to the leader of Germany, asking for advice on whether to declare war, and for Germany's military support. For some unknown reason, the delivery of the letter was delayed considerably, and the reply failed to arrive on time. Meanwhile, confident of military support from Germany and spurred by the public outrage for retaliation, the empire declared war, which quickly developed into World War I. In reality, the reply from the German leader had advised against the engagement. Had the emperor been told that the letter had not been delivered in a timely manner, the world's history would have been quite different.

- 4. Encoding effectiveness is limited by time: Encryption algorithms are generally defeated by utilizing computers to implement successive trial and error until the correct key is obtained. The higher the complexity of the algorithm, the greater the number of possibilities that must be examined, implying a longer elapse time until the encryption is finally broken. Thus, a key design characteristic of a secure network is the time interval T_1, where T_1 represents the duration over which the integrity of P_1 is assured, where possible. Although T_1 may be arbitrarily large, its value is subject to change as computers becomes faster and our knowledge of algorithms becomes more sophisticated.

- 5. Underlying intent of the packet transmission: Although protecting a user's packet is a primary function of network security, often, achieving the objective underlying the transmission of P_1 in the first place is an equally important function. There is compelling evidence, especially in light of characteristic 4, that the latter function is more important in today's highly dynamic, distributed world and promises to assume even greater importance in the future. Thus, a secure network may resort to creating multiple copies of P_1 and transmitting them via different routes to ensure that at least one of them reaches the target user in a timely manner, directly contradicting characteristic 1.

- 6. The need for additional communications and computational intelligence under characteristics 1 and 3: The successful mitigation of characteristics 1 and 3 may require coordination between the sender, receiver, and the network, which, in turn, may impose the need for additional message communication. To process these messages, clearly, the nodes underlying the sender and receiver must allocate additional computational resources. As an example, consider the transport protocol TCP, which is designed to guarantee delivery of a sender's message. It employs a complex suite of messages and is significantly slow.

- 7. The need for computational intelligence under characteristic 2: The need for computational intelligence to encrypt P_1 at the sender and decrypt it at the receiver is clear.

- 8. Challenges in meeting the computational intelligence needs of security: Fundamentally, network security consists in checking and verification that require computational intelligence. Characteristics 6 and 7 underscore this requirement. The key challenge in implementing them is explained as follows. To date, computational intelligence has come to us in material form, whether as human beings, electronic circuitry, or programmable computers. Science has not yet succeeded in designing computing engines that constitute pure energy. Consider material transport networks including inventory management and air cargo networks where a computer may be carried along with every material item being transported and the computer's computational intelligence may be utilized during the transported item's transit through the network. In contrast, in communication networks, P_1 is manifested in electromagnetic form and cannot carry with it a computing engine during its transport through the network. Consequently, all of its computational intelligence needs, including those stemming from its security needs, must be addressed by the limited number of "matter-based" nodes in the network. Since the number of packets transported in a network is, in general, orders of magnitude higher than the number of nodes, the competition for computational intelligence, concentrated solely at the limited number of nodes, is fierce. Network security will experience a profound evolution if and when science is able to synthesize computing engines in pure energy form.

- 9. The need to integrate security at the time of network design: From an examination of characteristics 6, 7, and 8, it follows that logically, security considerations must be incorporated at the time of network design, i.e., when the network is on the drawing board, not as an afterthought. The notion is similar to the design for testability in VLSI that gained tremendous momentum in the 1980s. For otherwise, the allocation of computational resources and the network control algorithm design that have been committed at the time of network design may require subsequent modification. This may be very difficult at best and impossible at worst. As an example, consider today's IP network, which was originally designed as the Arpanet in the 1970s as a store-and-forward network with at most a few hundred nodes, located within the US or at US military bases abroad, permitting computer communication between at most a few thousand users. Given its research nature, the original Arpanet designers focused on the success of computer-to-computer communications, beyond the contemporary voice communications over the telephone system. Intense security was neither warranted nor considered at the time of the network's design. The designers also did not foresee that the Arpanet would one day evolve into the Internet—a key vehicle of the world economy, consisting of hundreds of thousands of nodes, located all over the world, serving millions of users, and requiring secure communications. Today, Internet security has become a top priority, especially in the light of well-documented financial loses that have resulted from break-ins. However, hardening the Internet has proved to be extremely difficult, a key reason being the following. A basic characteristic of store-and-forward networks is that the individual packets of a user's message are transmitted independently to the target node, possibly along different routes, the precise knowledge of which may not be available to the sender or receiver. Conceivably, one or more packets of the message may unwittingly travel through nodes of the network where they may be intercepted by perpetrators and subject to different types of mischief. In response to the threat of attacks, a significant number of organizations have started to employ firewalls to separate their internal networks from the Internet. However, while the cost is tremendous loss of performance, estimated at over 90% [9], a recent finding that insider attacks constitute over 80% of security breaches clearly opposes the firewall philosophy. By design, ATM networks are more promising, in terms of security, but they are not totally immune from attacks.

- 10. Growing system complexity poses a challenge to network security:

$$O(t + 1) = f_1(I(t), S_i(t)),$$
$$S_o(t + 1) = f_2(I(t), S_i(t)), \qquad (1.1)$$

Unlike the small-scale systems of the past, today's networked systems are highly capable, powerful, useful, encompassing, far-reaching, and therefore complex. The systems of tomorrow promise to be even more complex. Under these circumstances, it may be impractical to expect a complete and precise characterization of such systems. Thus, in equations 1.1 for the output (O) and the subsequent internal state (S_o), expressed as functions of the external input (I), internal state (S_i), and time (t), f_1 and f_2 may not always be known with complete precision. Often, the best one can hope for is a combination of a partial characterization of the system and key insights into the most important elements of system behavior. Clearly, this leaves open the definite possibility for an imaginative and clever perpetrator to devise innovative attacks at any time during the life of the network to defeat the best attempts at hardening the network. Recognizing this, the secure network designer must continually assess the network for vulnerabilities and never fall prey to complacency relative to the security of the network. Consider a hypothetical example where tensions are running high between the US and nation X, and war is imminent. The final decision on whether to engage and the timing of the first air strike against X are clearly held top secret by the top US officials, and elaborate and expensive security resources are deployed to ensure secrecy. The news media and everyone else are waiting anxiously to learn of any new developments. At 6 pm on a specific day, a huge pizza truck pulls in front of the Pentagon, and observers outside immediately realize that the air strike will commence in a few hours. They reason that the food has been ordered in a hurry to feed the huge number of Pentagon employees who must stay behind all evening and into the night to support the logistics of the war. Had an informant of X been waiting outside along with the members of the news media, nation X would have gotten a valuable advance warning of the imminent attack by a few precious hours.

As a second example, consider the recent failure of a US military satellite in detecting signs of preparation leading up to an underground thermonuclear test in nation Y. Given widespread rumors about the imminent testing in Y, US military analysts must have been engaged in meticulously examining the spy satellite (SS) photos every time it passed over nation Y. Yet, no advanced warnings were issued until the actual explosion. It was later revealed that the military scientists of nation Y were tracking SS closely, utilizing their own satellites, and generated time intervals when SS was over other regions of the world. These intervals provided windows of opportunity for the military of nation Y to undertake the preparation for the thermonuclear testing, under total immunity from detection.

- 11. A fundamental vulnerability in every electronic-switch-based network: In any network, the traffic sources' interactions with the network elements are inherently asynchronous. That is, cells are inserted at a network node, arriving irregularly in time from the traffic sources. While the nodes of a network also interact asynchronously among themselves, faults and errors in the interactions between nodes may also occur asynchronously. In the most general case, the timing of the communication between two or more users on the network is likely to be asynchronous. If A and B are two users, neither A nor B can know, a priori and with certainty, when the other will initiate communication. Nevertheless, either one of them must be prepared to respond to the other when contacted. Also, when A contacts B, it has no a priori certain knowledge of how quickly B will respond. The synchronization of asynchronous processes is generally achieved by the use of synchronizers. Sophisticated synchronizers utilized in communication between two or more systems operating asynchronously are designed around the basic flip-flop, the most fundamental synchronizer in digital electronics. Under these circumstances, in theory [10][11][12], it is not possible to guarantee that the device specifications, setup and hold, will be satisfied by the input signals. When setup and hold time constraints are violated, the flip-flop may succumb to metastability, resulting in malfunctions that are transient, hard to correct, and occasionally devastating. Virtually every one of today's networks employs electronic switches. Given that flip-flops constitute an indispensable element of these switches, all of today's networks are vulnerable to the threat of metastability.

- 12. Analyzing the absence of security considerations in traditional network design: The sparse practice of integrating security considerations at the time of network design may be traced to two principal reasons. First, network design in itself is highly challenging, and the primary focus is invariably on successfully completing the design, economically and timely. In the industry the motivation to complete a design and start manufacturing it quickly is so strong that, often, the system design is specified only partially. That is, of the total number of possible input combinations, say N, to any one of which the system design may be exposed, the design is guaranteed to yield a specified behavior for only a subset, say M, of the N input combinations. Thus, the M input combinations adequately address the desired functionality of the system design. Although the manufacturer warranties correct system behavior only for these M input combinations and declares the remaining combinations (N – M) illegal, how the system will react to the (N – M) combinations remains unspecified and unknown. Clearly, these input combinations will continue to offer new opportunities for clever perpetrators to exploit them to launch attacks.

Second, the effort to incorporate security into a network design is undeniably hard, at least an order of magnitude higher in difficulty than the original design exercise. To gain a better appreciation of the difficulty of secure network design, it may be worthwhile to review the challenges faced by the designers of commercial aircraft, bridges, dams, passenger railways, and spacecraft in trying to prevent catastrophic failures. In a sense, the consequence of failing to defeat a security attack on a national medical records network in the future, leading to the release of personal medical histories of US citizens, may be no different, in human terms, than the untimely deaths of the crew of the space shuttle *Challenger*. The traditional approach to reliable design has generally required one to first enumerate the principal failure modes and then consider their impact, one at a time. The principal argument has been that the probability of multiple simultaneous failures is very low, that their prevention is excessively expensive, and thus they need not be considered. Analysis of the *Apollo 13* incident underscores this argument. A total of three simple unconnected engineering failures occurred simultaneously. A heater thermostat supposed to have been upgraded to accept 65 volts was overlooked and burned out at 28 volts. Several years before the flight, an oxygen tank had been dropped about 2 inches, further damaging the attached thermostat. When the astronauts followed orders to stir the tank, a routine procedure, the already damaged wiring melted, shorting out and causing a spark in an environment of pure oxygen. The resulting explosion blew the side of the spacecraft off. In reality, the crew had been exposed to each of these failures individually in the training simulator prior to their mission. When the astronauts were later asked how they would have reacted had they been exposed to all three failures at the same time in the simulator, they replied that they would have walked off the program complaining that the scenario was so far out and the probability so minuscule that there was absolutely no point in going through the exercise. Unfortunately, such failures do occur and are frequently associated with irreparable losses. In the network security discipline, the stakes are very high, promising to climb even higher in the future, and the demand for utmost precision in security considerations is real.

For example, the rapid proliferation of NIT systems in society, the growing concern among law enforcement agencies—including the US Department of Justice and the FBI—over how to prosecute increasingly sophisticated NIT related crimes, and the socially accepted strict legal standard encapsulated in "beyond reasonable doubt" demand that precision be incorporated in every phase of NIT systems design. Conceivably, spurred by law enforcement, legislators may enact laws mandating precision in NIT systems.

This book proposes a new approach that is both logical and practically realizable. It requires a designer of a secure network to analyze the key vulnerabilities of a network from first principles by acquiring a fundamental and comprehensive understanding of the underlying network, then synthesizing attack models to forcibly expose the vulnerabilities, and finally testing the models through modeling and asynchronous distributed simulation that closely resemble reality.

1.7 Scientific Validation of Secure Network Designs Through Modeling and Simulation

The traditional approach to understanding the behavior of real-world networks and, to a lesser extent, network security has been to develop analytical models that attempt to capture the system behavior through exact equations and then solve them using mathematical techniques. This has been adequate in the past and may continue to serve effectively in many disciplines. However, for many of today's networked systems, given their increasing size and complexity, which implies a large number of variables and parameters that characterize a system, wide variation in their values, and great diversity in the behaviors, the results of analytical efforts have been limited. Continued reports of perpetrators successfully penetrating the Internet and the lack of a conceptual solution in the literature underscores the limitation of the traditional approach. Tomorrow's networked systems are expected to be far more complex and far-reaching, implying that modeling and large-scale simulation may be the most logical and, often, the only mechanism to study them objectively.

Modeling refers to the representation of a system in a computer-executable form. The fundamental goal is to represent in a host computer a replica of the target secure networked system architecture, including all of its constituent components, as accurately and faithfully as possible. Simulation refers to the execution of the model of the target system design on the host computer, under given input stimuli, and the collection and analysis of the simulation results. The benefits of modeling and simulation are many. First, they enable one to detect design errors, prior to developing a prototype, in a cost-effective manner. Second, simulation of system operations may identify potential problems, including rare and otherwise elusive ones, during operation. Third, analysis of simulation results may yield performance estimates of the target system architecture. In contrast to the past, the increased speed and precision of today's computers promises the development of high-fidelity models of secure networked systems, ones that yield reasonably accurate results quickly. This, in turn, would permit system architects to study the performance impact over a wide variation of the key parameters, quickly and, in a few cases, even in real time or faster than real time. Thus, a qualitative improvement in system design may be achieved. In many cases, unexpected variations in external stress may be simulated quickly to yield appropriate system parameter values, which are then adopted into the system to enable it to successfully counteract the external stress. Last, the design of new performance metrics may be facilitated to gain a better understanding of the nature of the system behavior. The issue of determining appropriate input traffic demand patterns is discussed in detail in Chapters 4 through 7.

1.8   Problems and Exercises

1. Find out the earliest known record of the use of security in society. Was it in the military? What was its nature?

2. Read the early history and development of the Enigma machine used during World War II. What motivated its development and why was it not available during World War I?

3. Research into the techniques used by the Allied codebreakers to decipher the Enigma machine.

4. Obtain and read *The American Black Chamber*, by Herbert O. Yardley, and determine how the code breaking teams in different nations functioned during World War I.

5. Obtain a copy of the documentary film *Extreme Machines Trident Submarine*, aired by The Learning Channel (TLC), September 20, 2000, and analyze the submarine command and control structure.

6. Even after they were provided with the details of the communications scripts between headquarters and the U-boats and the timings and locations of the sinkings of several U-boats, the military scientists in Germany during World War II concluded that the Enigma code had not been broken by the Allies. Find out the reasons.

2
A Fundamental Framework
for Network Security

2.1 Introduction

A natural starting point in implementing a network security system should consist in a comprehensive definition that includes all areas related to network security and applies to all types of users from the military, government, and industry. Extensive search reveals the lack of such a definition or framework in the literature, and the underlying reason may be described as follows. Different classes of users have developed their own unique definitions to encapsulate their own security concerns, and their frameworks are incompatible with one another. While these unique definitions may have been adequate when networks were closed and isolated, they are inappropriate in today's climate of increased interconnection between networks. Without a common definition for network security, users can no longer protect their data in interconnected networks. The need for a standard definition is genuine, and it must enable a unified and comprehensive view of security among civilian, military, and government networks. It must provide a basis to address, fundamentally, every weakness in a given network. It must also apply to every level of the network, starting at the highest network-of-networks level and descending to the single computing node that maintains connections with other nodes. In essence, the common standard for defining network security will enable the understanding of the security posture of an individual network, comprehensively facilitate the comparative evaluation of the security of two or more networks, and permit the determination of the resulting security of a composite network formed from connecting two or more networks. It is important to observe that the framework for network security constitutes a methodology for organizing and categorizing actual implementations of network security. The framework does not provide implementations of network security. Instead, it offers a map for organizing and describing mechanisms to achieve practical network security. Consider, for example, a specific encryption device that can both encrypt and decrypt data on a communications link. While the device corresponds to an implementation of network security, the specific security area constitutes communications security. For further details of security devices the reader is referred to Stallings [13], Pfleeger [14], and White, Fisch, and Pooch [15].

2.2   The Changing Paradigm of Network Security

In the last decade, threats to network security have evolved, resulting in a dramatic change in the manner in which network security is viewed. In contrast to the classic threats including the military espionage of the former Soviet Union, hostile governments, and terrorist organizations, the new threats include information warfare and, more recently, liability law suits stemming from the failure to protect information. While there are always threats from the outside, Edfors [16], Madron [17], and Simonds [18] reveal that the greatest threat has always been and continues to be "insiders." Madron estimates that as high as 75% of possible attacks on a network emanate from one or more sources within the network establishment. In response, both government and industry are emphasizing internal threats. Threats may be leveled at a network from several different aspects of the network. Thus, while one hacker may infiltrate an application and divert financial transactions to a fictitious account, another intruder may attack the performance of the network and render it virtually unusable. The following is a partial list of the most important sources of threats.

- Outside threats: Terrorism, hackers, malfeasants, former employees, foreign political espionage, economic espionage by foreign governments or foreign or domestic corporations, and liability lawsuits.

- Inside threats: Employees, hackers, mischief, legal, and uneducated users.

The literature records the use of audit trails, cryptography, and authentication techniques to ensure network security. However, these techniques are essentially confined at the lower level and are ad hoc. Hitchings [19] strongly believes that a new approach to information security is needed. This chapter theorizes on the need for a logical, high-level approach to comprehensive network security and argues that the networks and algorithms underlying high-level applications must integrate security concerns into their design. The overall approach must continuously monitor the exchange of data between sites and within a site to detect or defeat intrusions and unauthorized activities as well as a fail-safe method for detecting system faults before they become catastrophic. The approach must anticipate the presence of malicious and ignorant users everywhere and not take the validity of any data for granted.

Increasingly, experts in computer systems are recognizing the vital role of system availability or stability, a concept that has long been recognized in the electric power community. If an enemy succeeds in degrading the command and control network sufficiently, none of the sophisticated cryptographic schemes are useful, since no messages will get across the networks. The nation's well-being is at risk if an enemy is capable of rendering the telecommunications network unavailable without engaging the traditional defense forces or firing a single shot.

The phenomenal growth of distributed networked systems, coupled with their enormous future potential, marks a new era in the information age. On the other hand, their ubiquity and the near total dependence for day-to-day activities, coupled with their complexity and vulnerability, raises a deep concern. The algorithms that underlie distributed systems are complex and are susceptible to external and environmental disturbances. The well-known incidents of unexplained system failures in IBM mainframes despite their having run smoothly for decades, the discovery of previously unknown errors in AT&T network software, and the numerous unreported telephone system failures across the country all attest to the complexity of the underlying algorithms. A researcher from a major manufacturer of automated teller machines acknowledges that the nation's network is a collection of patchwork and that it is a miracle that it works. The perturbations, even if transient, may degrade system performance, either lightly or severely, or even cause catastrophic failures. It is therefore imperative to undertake serious efforts to understand the issue of stability in depth and to synthesize performance metrics. These changing threats have caused a shift in the network security paradigm from one of certification to one of risk assessment.

Fundamentally, the underlying reason for network security is the value of the information riding on the network. According to Madron [17], Admiral Grace Hopper pointed out in the 1970s that the industry is engrossed only in the processing aspect of the information processors and lacks a basic understanding of the "value of the information." Even though computers and networks have been around for decades, there appears to be a lack of a community-wide agreement on adopting a framework to define, describe, and evaluate network security.

2.3 Review of the Literature on Network Security

In the literature, definitions of network security terms are influenced heavily by the respective researcher's affiliation and background: industry, government, or military. Each of the three sectors continues to maintain its individual vocabulary, which is built around the perceived threat and cost benefit. However, the distinction is increasingly being blurred by overlapping networks, as is highlighted by a recent fact: 90% of the DoD's electronic traffic runs over the public networks [20]. The lack of a common language to describe network security and the consequent inability to discuss network security hampers progress in the field and threatens the livelihood of millions of people and hundreds of corporations and government agencies. It is therefore imperative for all involved parties to agree on a common framework and revitalize efforts toward evaluating network security.

The development of security in automation and control over the years has been ad hoc, led primarily by the available technology and the goals of the funding agency. During World War II, the focus was on cryptography, which aimed to protect written traffic between encoding and decoding machines. This is defined as communications security. With the proliferation of computers and the birth of networks, the role of cryptography expanded into military networks. However, as discussed earlier, cryptography is only one attribute of a secure network and it alone cannot guarantee comprehensive security, particularly with today's and tomorrow's sophisticated computer-literate population.

The US military methodically categorizes security attributes in the Orange Book [21] and the Red Book [22]. While the concepts of COMSEC and INFOSEC are well understood within the Department of Defense, they mean little to most of industry and many civilian government agencies. A comprehensive literature search was carried out that culminated in a detailed listing of the attributes of network security, as used by the military, and the specific terms used to describe them in industry and government. Common terms were grouped together, and the best term was selected to describe the respective issue. As an example, consider the term "classification," which the military uses to describe whether a network is restricted to a particular person, group, or class. This is further subdivided into the categories of unclassified, for official use only, confidential, secret, and top secret. In contrast, the term "private" or "proprietary" in industry restricts the use of a network to a specific person, group, or class and therefore corresponds to the military's "classification." Motorola's POPI classification is based on whether failure to protect data may disrupt business, provide undue economic advantage to the receivers, cause embarrassment, permit access to other classified data, provide undue advantage to a competitor in the marketplace or in its negotiations with a mutual customer or in its market strategy or access to technology, or lead to legal problems including liability.

Schwartau [23] cites a number of sources in the literature that propose definitions of network security and enumerate lists of network security attributes. Schwartau [23] even lists a number of ways to subvert network security. The definitions include protecting the originality of the data, encrypting data, reliability of the network, privacy of data, physical security of the network and its links. None of the definitions, however, is comprehensive, and the lists of attack mechanisms jump from the DoD's perspective to that of other government agencies and industry. Simonds [18] cites several examples of security standards including the DoD's Orange Book, OSI security services by ISO model layer, as well as listings of areas of network security such as authentication, access control, and encryption. These standards and areas of network security are neither comprehensive nor self-contained, since they are presented from the viewpoints of the different sectors. Winkler [24] presents numerous examples of how security is routinely violated in the corporate arena but, again, does not provide a comprehensive definition of network security. Efforts such as the Common Criteria [25] have focused on combining many security attributes into one unified standard but have failed to defend the standard as comprehensive.

A key difference between the industry, government, and defense perspectives has traditionally hovered around threats and their sources. This difference is also becoming blurred. To the military, the traditional threat has been an enemy, typically a hostile government or terrorist, whose efforts were aimed at stealing valuable information from the network. Today, however, a new kind of threat, termed "information warfare," is gaining notoriety. It consists in disabling or rendering useless the enemy's key networks including the command and control [26], power grid [20], financial, and telecommunications networks. In addition, the threat of economic espionage, i.e., stealing secrets from industry and government networks, is on the increase. There is increasing evidence that insiders—disgruntled and recently fired employees—constitute the most significant threat [16]. Malefactors are another threat capable of causing mischief or serious harm to networks.

In general, the definitions of network security in the literature are neither comprehensive nor universal. They are usually focused on the particular class of users perceived as a threat to their networks and security measures employed in the past. These definitions, from different classes of users, military, government, and industry, are usually in the form of lists and in numerous cases use different vocabularies to describe the same aspect of network security. None of the definitions incorporates all of the important aspects of network security described in the literature.

Madron [17] presents a generalization of the DoD's network security vocabulary of such terms as INFOSEC and COMSEC and provides the following definitions. INFOSEC is defined to consist of procedures and actions designed to prevent, for a given level of certainty, the unauthorized disclosure, transfer, modification, or destruction, whether accidental or intentional, of information in a network. Information includes data, control, voice, video, images, and fax. In contrast, COMSEC refers to protection resulting from the application of cryptosecurity, transmission security, and emissions security measures to telecommunications and from the application of physical security measures to communications security information.

According to the literature, a number of authors have provided additional vocabulary to describe aspects of network security. The following is a review of publications found during the literature search in this field. Abrams and Joyce [27] review trusted-system concepts and reference validation mechanism and explore a new computer architecture to generalize the concepts for distributed systems. Nessett [28] reviews the difficulties in authentication and notes the security advantages of centralized authentication during logon in distributed systems. Lin and Lin [29] note that in enterprise networks, the principal security "areas" include confidentiality, integrity, data-origin authentication, nonrepudiation, user authentication, and access control. They review public-key and secret-key cryptographic techniques for confidentiality and Kerberos for third-party authentication. They also suggest the use of centralized security management over distributed schemes to reduce overhead and security risks. Cryptography has continued to play a major role in security. To Janson and Molva [30], network security involves the tasks of controlling access to objects, enumerating the access rights of subjects, the threats that must be considered during access control design, and mechanisms to enforce access control. They describe the role of cryptography as central to both authentication and access control. In addition, they propose tracking resource usage by authorized users at least for accountability and for subscribers to identify themselves to each other to fend off masquerading intruders.

Power [26] introduces the notion of information warfare and notes that its scope includes (i) the electronic battlefield, i.e., disruption of enemy command and control, (ii) infrastructure attacks, i.e., failing key telecommunications, financial systems, and transportation, (iii) industrial espionage, i.e., covert operations aimed at stealing proprietary secrets or sabotage of company information networks, and

(iv) personal privacy, i.e., stolen private information such as credit card or driver's license or social security numbers. The security services required in electronic commerce (EC) networks [31] include authentication, authorization, accountability, integrity, confidentiality, and nonrepudiation. Geer [31] also identifies two kinds of possible attacks on EC networks: (i) passive, or pure listening, and (ii) active, or insertion of modified packets. To defeat such attacks, the goals of security must be aimed at preventing traffic analysis attacks, preventing release of contents attacks, detection of message stream modification attacks, detection of denial of service attacks, and detection of spurious association initiation attacks. Hosmer [32] remarks that the desired goal in the current computer security paradigm is absolute security. This requires logical and mathematical precision, while unfortunately, precision and complexity are inversely related. A related complication is that the future may witness new types of threats to network security.

According to Hill and Smith [33], the risks in the corporate world include personnel, property, information, and liability. Today's corporations are concerned with (i) protecting financial resources, personnel, facilities, and information, (ii) access control for facilities and management information systems, and (iii) recovery from disaster and continuity of operations. Chambers [34] underscores the difficulty in detecting intrusion and notes that although the FDA network was successfully penetrated in 1991, the logging and monitoring tools, left running for weeks, revealed no signs of unauthorized access. Wolfe [35] underscores the value of the information contained in the hardware by pointing out that for many likely events that arise from the lack of security, such as virus attacks, there is neither a widely accepted measure of risk nor is it likely to obtain insurance for information. Oliver [36] traces the concept of "privacy" of computer users and individual-related data to the US Constitution and notes that it is provided by a third party as far as distribution, publication, and linkage of information to the individual are concerned. Oliver also addresses the debate as to whether computer users making anonymous statements may be held accountable. Hitchings [19] stresses the need to examine the human issues: cultures of people involved, attitudes, morale, and differences between personnel and organization objectives relative to network security.

The literature on the use of audit trails to realize accountability, detect anomalous behavior of users, and possibly flag intrusion is rich. Vaccaro and Liepins [37] describe their experiences with recording and analyzing anomalous behavior in computer systems at Los Alamos National Laboratory immediately following an intrusion. Helman and Liepins [38] present a stochastic foundation for and analysis of audit trail analysis. They also suggest several criteria for selecting attributes. Janson and Molva [30] propose the tracking of system resource usage by authorized users for accounting as well as intruder detection. They mention the need to (i) identify objects access to which must be controlled, (ii) identify subjects whose access must be controlled, (iii) identify possible threats that must be defeated, and (iv) catalogue enforcement mechanisms. Lunt and Jagannathan [39] enumerate several discrete and continuous intrusion detection criteria and state that their system maintains system usage profiles of users, which in turn are periodically updated based on a priori known user behavior. Kumar and Spafford [40] encode the knowledge of known attack procedures through specialized graphs in their system and use of a pattern-matching scheme to detect network penetration. Soh and Dillon [41] present a Markov model of intrusion detection and devise a "Secure Computation Index" measure to quantify the intrusion resistance of a system. Their results, however, are limited to a single computer system. In her survey of intrusion detection techniques, Lunt [42] notes that they are primarily based on maintaining audit trails and observes a few key controversial issues. They include the appropriate level of auditing, the voluminous amount of audit information, the comprehensibility of detailed audit information, the possible performance degradation as a result of audit, and the invasion of privacy of computer users. A variation of the audit trail concept has been proposed for the electric power industry. Weerasooriya and colleagues [43] present a neural-network solution to the problem of security assessment in large-scale power systems. They use neural nets for fast pattern matching of the state of the power system immediately following a "contingency" with historical trends. Their results are, however, limited to static security. Recent research in intrusion detection continues to focus on the use of audit trails [44] [45], attempt to detect patterns in the traces of data and privilege flows [46][47], and employ statistical and neural-network models [48][49][50]. Midori [51] propose the use of autonomous agents to collect break-in information. Following testing of the current intrusion-detection products by vendors, Newman, Giorgis, and Yavari-Issalou [52] observe that no product is capable of successfully detecting all attacks under heavy network loading, a conclusion corroborated by Lunt [53].

Recently, many computer network experts [16][20] have joined electric power system researchers in sharing the latter group's long-held belief in system availability [54][55][56] and transient stability [57] as primary security concerns. Fitzpatrick and Hargaden [58] argue that the design of complex networks must take into account scenarios where the network may be rendered unavailable by enemy action. They point out that in military command and control networks, units may need to continue fighting while out of contact with higher headquarters and adjacent units, acting on their own initiative within the framework of the commander's intent.

An analysis of the current literature reveals the following. First, the nature of security concerns differs for each of the sectors: military, government, and industry. This has led to problems, since many of these sectors are forced, for reasons of efficiency and economy, to use each others' networks. Third, there is a lack of a common framework and vocabulary to describe security and intrusion resistance of networks, an important issue that dominated the Network Rating Model workshop. Fourth, traditional security criteria have already been transcended, and it is critically important to address the issues of stability of networks, intrusion resistance, privacy, and other high-level security issues.

2.4   The National Security Agency's Network Rating Model (NRM)

Concerned by the growing threat posed to the nation's key networks, the National Security Agency (NSA), the nation's chief proponent of national security, organized the first public workshop in March 1996 and two subsequent Authors' Group meetings in 1996 and 1997. NSA's objective was to develop a Network Rating Model (NRM), through consensus, that would be acceptable to the military, government, industry, and academia. The NSA expected the NRM to encapsulate its vast expertise in security products and certification. The NRM would include a definition of network security, identify potential threats, determine the degree of protection that should or could be provided to a network, synthesize a measure of protection and a methodology for evaluation, and determine cost and performance tradeoffs. Logically, one must bound what one is protecting before one can analyze the effectiveness of the protection. Thus, the attributes serve as potential weak points in a network. It is clear that the vulnerability or security of a network may be viewed from different conceptual points of view, termed perspectives by the author. Although this idea is referred to as "disciplines" in the literature, the term "perspectives" appears to capture the underlying meaning more accurately. Therefore, the total security of a network requires its detailed evaluation, relative to every perspective. While one organization, building on its assumption of a specific set of threats, may find one subset of the perspectives important, another organization may find a different subset of the perspectives critical based on its own perceived threats.

The consensus definition of a NRM is this: "a consistent, cost-effective methodology based upon a defined set of characteristics for assessing the total security of any network or combinations of networks, either in operation or development; to define what exists, determine what is needed, identify what could affect security, and provide a universally acceptable assessment report."

In the definition, the term "consistency" stresses the need for the security rating of a network to apply uniformly across different sectors. Furthermore, a rating must be valid for a reasonable length of time into the future despite rapid advances in networking technology. The cost effectiveness criterion underscores the need to balance the cost of the threat against the cost of implementing security. Total security refers to the different dimensions of a secure network, while the phrase "network or combinations of networks" reflects the increasing blur between network boundaries. Since the effort must be universally acceptable and useful, it must record the security measures currently in place in the networks which, in turn, will facilitate identifying what more is required to ensure total security.

In order to define the characteristics or attributes of a given secure network, it was agreed at the workshop that one must focus on the relevant set of network security perspectives to yield security services that satisfy stated concerns. The comprehensive list of perspectives includes (a) systemic, (b) communication, (c) physical, (d) personnel, (e) operational, (f) application, and (g) performance. The services were enumerated as (a) access control, (b) confidentiality, (c) integrity, (d) authentication, (e) traffic flow security, (f) assured service, (g) nonrepudiation, (h) anonymity, and (i) intrusion detection. The concerns included (i) accountability, (ii) availability, (iii) liability, (iv) reliability, (v) audit-ability, (vi) interoperability, (vii) confidentiality, and (viii) integrity. These perspectives, services, and concerns are corroborated in [16].

At the first NRM workshop, given the limited time available for a thorough discussion, security services and concerns were separated into two distinct lists. This split was driven by the divergent views of the representatives of industry, government, and military, which, in turn, stemmed from differing perceptions of the sources of threats. Following the conclusion of the workshop and upon careful analysis, it became increasingly evident to the author and his research team that a unified approach to total network security, i.e., across the military, government, industry, and university sectors, requires the recognition of two fundamental components of network security. First, any secure network must possess a few inherent characteristics, regardless of the sector to which it belongs and independent of any specific threat. The characteristics are referred to as attributes of a secure network and they are the result of unifying security services and concerns. Second, a network's security may be viewed at different conceptual layers, each view reflecting a threat, relatively orthogonal to others and thereby permitting independent development and evaluation.

The final NRM report was presented by NSA at the twentieth National Information System Security Conference in Baltimore in October 1997. It incorporates the author's fundamental framework for network security and is crucial because it validates the need for a common framework across all sectors: military, government, industry, and academia.

2.5 A Fundamental Framework for Network Security

A framework is defined [59][60] as a conceptual structure that encapsulates the fundamental knowledge and the set of relationships of a discipline. A framework permits systematic and scientific reasoning about the discipline and is therefore essential to the advancement of the discipline. In the context of network security, the framework must be comprehensive, especially since the nature of networks is changing from a set of interconnected entities, controlled and used by a specific class of users, to an increasingly interconnected and integrated network that is simultaneously shared by different classes of users and utilizes a language that is common to industry, government, academia, and the military.

Figure 2.1 presents a representation of the author's framework by means of a matrix, or two-dimensional array, where the two axes, labeled pillars and attributes, reflect two orthogonal views of network security. The pillars represent the constituent components of the network that collectively support the security posture of the overall network and are subject to threats. For example, the physical pillar refers to the physical enclosure surrounding a network element including guard doors and locks and is subject to threats. Each of the pillars is amenable to independent development and evaluation, and furthermore, the strength or weakness

Network Security Attributes

Network Security Pillars	Privacy	Integrity	Accountability	Availability	Reliability	Connectivity	Recovery	Liability	Uncertainty
Systemic									
Communication									
Physical									
Personnel									
Operational									
Application									
Performance									
Design Correctness									

Figure 2.1 Comprehensive network security framework.

of one pillar may not be transferred directly to any other pillar. The attributes refer to the one or more inherent properties that must characterize any secure network and must be defined independent of any threat. Thus, the attribute "recovery from disaster" refers to the inherent property of resilience of the network to perturbations and must be defined independent of the threats. For a given network, the two-dimensional matrix representation of the framework assumes that the pillars and attributes are arranged in rows and columns, respectively. The values of the matrix elements along a column relate to the pillars, each of which contributes to the strength of the corresponding attribute.

Conceptually, a secure network may be viewed as one where the attributes permeate each of the pillars, which, in turn collectively hold up network security. The relative strengths of the pillars may vary, depending on the perceived threats in a given scenario. Thus, network security is only as strong as the weakest pillar. The concept provides an organized framework for the network security evaluation information, which may be utilized to improve security or to evaluate the resulting security from interconnecting two or more networks. Ideally, a fully secure network would require every attribute to be strongly protected in all pillars, subject to some standard threat, relative or absolute. However, this may be neither cost-effective nor practical due to limited time and resources. Decisions related to network security are based on the perceived threat to a particular pillar and/or attribute and the level of risk that the security management is willing to assume.

The list of attributes includes (1) privacy, (2) integrity, (3) accountability, (4) availability, (5) reliability, (6) connectivity, (7) recovery from disaster, (8) liability, and (9) uncertainty, and they constitute a superset of the attributes proposed in the literature. The list of pillars include (a) systemic, (b) communication, (c) physical, (d) personnel, (e) operational, (f) application, (g) performance, and (h) design correctness.

2.5.1 Pillars of Network Security

The choice of the term "pillars" reflects the eight foundation blocks, each of which may be under attack, either independently or collectively. They cumulatively support a network's security. Thus, the pillars corresponding to the eight perspectives reflect the eight foundation blocks that individually describe an orthogonal conceptual view of network security and may be developed and evaluated independently, based on the degree of importance assigned to the appropriate threats. Consequently, the pillars may exhibit different relative strengths. Orthogonality implies that no matter how much improvement is incorporated into one pillar, it cannot compensate for the weakness in another pillar. Should new types of threats emerge in the future, requiring additional views of network vulnerability, additional pillars may need to be incorporated into the framework. The scope of the eight pillars is elaborated as follows:

- *Systemic* encompasses the software that operates the network and constitutes the basic infrastructure of the high-level application software.

- *Communications* encompasses the links and devices that interconnect the computers to constitute the network.

- *Physical* encompasses the equipment, material, and documents associated with the network.

- *Personnel* encompasses the people associated with the operation or use of the network.

- *Operational* encompasses the procedures, policies, and guidelines that constitute the security posture of the network.

- *Application* encompasses the high-level software that is executed on the network.

- *Performance* encompasses the normal range of operating parameters and throughput of the network.

- *Design correctness* encompasses the correctness of the total system. The complex interactions among the different components of the system will, in general, result in a very large number of states and state transitions. Without ensuring that every state and state transition is correct, the threat of the system entering an unstable state that then triggers catastrophic failure is very real. There is a much more important and fundamental issue that underscores the design correctness pillar, as described subsequently. Virtually every one of today's networks employs electronic switches. In turn, flip-flops constitute an indispensable element of these switches. Given that incoming signals are asynchronous, under these circumstances, in theory [61][10][11][12], it is not possible to guarantee that the flip-flops' device specifications—setup and hold—will be satisfied by the input signals. When setup and hold time constraints are violated, flip-flops may succumb to metastability, resulting in malfunctions that are transient, hard to correct, and occasionally devastating. Thus, according to our present understanding, all of today's networks are fundamentally vulnerable to the threat of metastability.

2.5.2 Attributes of Network Security

Each of the attributes of network security will bear a specific degree of relationship to each of the eight perspectives, or pillars, defined by the network and the current understanding of security attacks. While most of the relationships are readily understood, a few are unclear at the present time, while all are subject to evolution as our understanding of network security matures. For instance, the privacy attribute bears a strong relationship to the personnel pillar. In contrast, consider the relationship between the performance pillar and the liability attribute. At the present time, the relationship is weak since it is difficult to prosecute a hacker for degrading a network's performance and even more difficult to quantify the degradation and therefore determine a commensurate punishment. However, as society acquires a better understanding of the responsibilities and consequences, the relationship will be greatly refined. The relationships may be evaluated, objectively or subjectively, through mechanisms, some of which are well known, while others are yet undefined. As an example, the use of background checks may help strengthen the privacy attribute and the personnel pillar. Similarly, the strength of the relationship between the systemic pillar and privacy attribute for a given network may be evaluated through the access controls implemented. While the dependencies between the (i) "design correctness" pillar and the attributes and the (ii) "uncertainty" and "liability" attributes and the pillars, are clear, the exact relationships and the corresponding mechanisms to evaluate them are yet to be defined. The attributes are elaborated as follows:

- Privacy [26][36] is defined as intended for or restricted to the use of a particular person, group, or class. It applies to data, control signals, and traffic flow. Synonymous and associated words in the literature include confidentiality [30][31], anonymity [36], classification [21], proprietary, TRANSEC, cryptosecurity, EMSEC, and encryption [17].

- Integrity [56][31] is defined as ensuring that information held in a system be a proper representation of the information intended and that it not have been modified, created, destroyed, or inserted by an unauthorized entity. Integrity also refers to processes, process sequences, and other system assets. Synonyms and associated words include soundness, incorruptibility, completeness, and honesty.

- Accountability [36] is defined as a statement or exposition of reasons, causes, or motives to furnish a justifying analysis or explanation that can be documented or traced and ownership established. Synonyms and associated words include nonrepudiation [56], audit-ability [42], audit trail [42], answerable, authentication [31], signature, and responsibility.

- Availability [30][56] is defined as qualified and present or ready for immediate use by authorized users and worthy of acceptance or belief as conforming to fact or reality. Synonyms and associated words include access control [30], authentication [31], and confirmation.

- Reliability is defined as generating consistent results during successive trials. Synonyms and associated words include assured service, assuredness, certainty, and dependability.

- Connectivity [62] is defined to consist in the devices that constitute the network, including the computers and links between them, and the intelligence that supports the seamless and transparent integration of a wide variety of different protocol-driven terminals and host computers. Synonyms and associated words include interoperability, traffic flow, logical flow, associations, relationships, emissions control, and TEMPEST.

- Recovery [33] is defined as returning from a disaster and continuity of operations. Synonyms and associated words include self healing and contingency planning.

- Liability [33] is defined as having to do with legal obligation and responsibility that may affect property and information. Synonyms and associated words include responsibility, due process, ethical responsibility, open, and exposure, i.e., lack of protection or powers of resistance against something actually present or threatening.

- Uncertainty reflects the lack of complete knowledge of system security as a result of previous penetrations, with known and unknown consequences, that may degrade future network security. A perpetrator may leave behind "Trojan horses," hidden deeply in user data or programs and lying dormant. Even where the system hardware is completely reinitialized and the operating system and other application software reloaded onto the hardware, the dormant code may be triggered, either inadvertently by user activity or intentionally by the perpetrator, resulting in damages. This attribute may be viewed as a generalization of the concept of anomaly detection [39] in a user's behavior through audit trail analysis.

2.6 Uses of the Fundamental Framework for Network Security

The framework provides a basis to address, fundamentally, every weakness in a given network. Furthermore, it applies to every level of the network, starting at the highest network-of-networks level and descending to the single computing node that maintains connections with other nodes. Thus, the framework enables the understanding of the security posture of an individual network in a comprehensive manner, the comparative evaluation of the security of two or more networks, and the determination of the resulting security of a composite network that is formed from connecting two or more networks with known security.

The value of the proposed framework is that it stimulates network designers to examine the vulnerabilities of all eight pillars even when they may appear inconsequential. This constitutes the first use of the framework. For instance, while a credit card network operating on the Internet may successfully address the privacy attribute and feel secure, malicious agents may penetrate the network and reduce availability such that customers are prevented from making purchases. An examination of the performance pillar may be advisable under these circumstances. In a different scenario, while the military assigns resources to ensure the privacy and connectivity attributes, a disgruntled employee may send an unauthorized message, under an assumed identification, to the finance and accounting military pay program, take advantage of a weakness in the accountability attribute, and deny hundreds of thousands of soldiers their pay on time.

The procedure for determining a rating for a given network consists of the following. For a given standard threat level, relative or absolute, and a given environment, the strengths of the intersection points in the matrix are obtained through evaluating the corresponding mechanisms. The evaluations may assume the form of numerical values, narratives, or graphs, subjective or objective. To improve the security posture of the network, either (i) the individual values along a row that constitute an evaluation of the corresponding pillar may be examined against a perceived threat level, or (ii) the values along a column that reflect an evaluation of the strength of the corresponding attribute may be compared against a desired measure for the attribute. Clearly, the desired measure will reflect a cost-benefit analysis, i.e., the level of risk that the security management is willing to assume. As indicated earlier, the matrix provides a meaningfully organized framework for the network security evaluation information, in terms of its fundamental characteristics. Thus, to compare the security postures of two or more networks, either (i) the individual values along a row of the corresponding matrices may be examined against each other, or (ii) the values along a column of the corresponding matrices may be contrasted.

To understand the operation of the framework, consider that the military perceives the primary threat to its networks and data from hostile governments. Clearly, to the military, the communications and physical pillars are vulnerable. This, in turn, points to the Connectivity attribute. Furthermore, the desire to protect data riding on the network requires focus on the privacy attribute. In contrast, consider a financial network's concern that a malicious agent may disrupt its financial services. Clearly, the systemic pillar is vulnerable, which in turn points to the connectivity attribute. In addition, the privacy attribute may also be flagged due to the confidential nature of the financial transactions.

Assume that a defense agency plans to send Top Secret traffic through an ATM network. Initially, it will insert the highest value, say 0, in the entire privacy column and communication row of the matrix associated with the corresponding call request. To successfully propagate the traffic through the network, the call setup process must first determine a route, if possible, where each and every ATM node along the route offers a privacy value of 0 in every pillar and 0 in every attribute of the communications pillar. The values assigned to the elements of the security framework matrix of the node reflect the strengths of the security in the respective domain. The values of the individual elements may differ over a wide range, with some elements possibly having the value nine, implying the absence of security in that element area. Examples of three matrices, corresponding to three types of military traffic - Top Secret, Secret, and Confidential - are presented in Figure 2.2 along with the relevant element values.

It should also be noted that network security is a continuous process and must be exercised periodically. With time and as the roles of networks evolve, security breaches may appear in previously unsuspected areas.

The framework's second use is in computing the resulting security of the composite network AB formed from connecting two networks A and B with known security and shown in Figure 2.3.

By design, the framework applies to every level of the network, starting at the highest network of networks level and descending to the single computing node that maintains connections with other nodes. A key goal of the framework is to provide a template for organizing the different aspects of network security to permit the military, government, and industry to start their connectivity discussions from a common baseline. Whether they choose to use or ignore some or all of the elements of the framework is their decision and is based on the amount of risk they wish to assume. In any case, they will all be aware of the total framework and all of its elements. The perspectives, termed pillars, individually provide orthogonal views of network security and collectively constitute a comprehensive stable structure that supports the total network security. The attributes refer to the inherent characteristics of a secure network.

There is a third and more exciting use of the framework in which it enables a true security-on-demand paradigm. This is the subject of discussion of the next chapter.

The importance of the security framework is underscored from yet another perspective. From practical considerations, it is unrealistic to expect any single organization to possess adequate resources to interconnect every important entity in the world. Therefore, connectivity among organizational networks is a natural evolution which, in turn, implies the importance of the security matrix.

For any given network, with each element of the security matrix one may also associate either a (i) weakness, manifested as a vulnerability or threat, or (ii) strength, expressed in the form of a robust mechanism. Filling in all of the elements may require significant effort and care. From perspective (i), the deployment of cryptography addresses the elements at the intersection of the "communication" pillar and "privacy" and "integrity" attributes, while Kerberos addresses the element at the intersection of the "communication" pillar and "accountability" attribute. The notion of security on demand, introduced subsequently in Chapter 3, addresses the elements at the intersection of the "performance" pillar and

Top Secret	Privacy	Integrity	Accountability	Availability	Reliability	Connectivity	Recovery	Liability	Uncertainty
Systemic	0								
Communication	0	0	0	0	0	0	0	0	0
Physical	0								
Personnel	0								
Operational	0								
Application	0								
Performance	0								
Design Correctness	0								

Secret	Privacy	Integrity	Accountability	Availability	Reliability	Connectivity	Recovery	Liability	Uncertainty
Systemic	3								
Communication	3	3	3	3	3	3	3	3	3
Physical	3								
Personnel	3								
Operational	3								
Application	3								
Performance	3								
Design Correctness	3								

Confidential	Privacy	Integrity	Accountability	Availability	Reliability	Connectivity	Recovery	Liability	Uncertainty
Systemic	5								
Communication	5	5	5	5	5	5	5	5	5
Physical	5								
Personnel	5								
Operational	5								
Application	5								
Performance	5								
Design Correctness	5								

Figure 2.2 User-specified security matrices for military traffic.

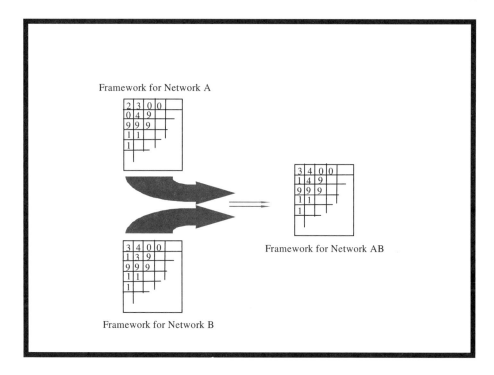

Figure 2.3 Comparing the security of two networks.

"privacy," "integrity," "availability," "reliability," and "connectivity" attributes. From perspective (ii), attacks 1 and 2, described subsequently in subsections 6.4.3 and 6.4.4, refer to the elements at the intersection of the "systemic" pillar and "privacy," "integrity," "accountability," and "availability" attributes. Attacks 3, 4, and 5, also elaborated subsequently in subsections 6.4.5, 6.4.6, and 6.4.7, address the elements at the intersection of the "application" pillar and six attributes extending from "privacy" through "connectivity" as well as those at the intersection of the "performance" pillar and "availability" and "connectivity" attributes. The complex attack detailed subsequently in section 7.1 refers to the elements at the intersection of the "performance" pillar and "availability" and "reliability" attributes.

2.7   Problems and Exercises

1. Provide real-world scenarios that highlight the "design correctness" pillar.

2. Given a network with a high degree of the "uncertainty" attribute, stemming from previously successful attacks, what techniques can you suggest through which the problem may be (i) mitigated? (ii) totally eliminated?

3. Which elements of the composite fundamental security framework F_{AB} resulting from combining two networks N_A and N_B, with the security frameworks labeled F_A and F_B, respectively, are most likely to be problematic? Assume (i) that F_A is a banking network and F_B a military network, (ii) that F_A is a patient medical records network and F_B a major corporation's network.

4. For your organization's network, identify the key elements of the security matrix and suggest mechanisms to assess their qualitative or quantitative values.

3
User-Level Security on Demand in ATM Networks: A New Paradigm

3.1 Review of the Literature on Security in ATM Networks

Since World War II, the focus in the security community has been on cryptography that aims to protect written traffic through encoding and decoding. With the proliferation of computers and the birth of IP networks, of which the Internet is a prime example, the role of cryptography has also expanded and has continued to dominate network security. Security in the Internet assumes the form of encoding data packets through cryptographic techniques [63] [64] coupled with peer-level, end-to-end authentication mechanisms [65], such as Kerberos [66], at the transport or higher layers of the OSI model. This is necessitated by a fundamental characteristic of store-and-forward networks: that the actual intermediate nodes through which packets propagate are unknown a priori. A potential weakness of this approach may be described as follows. Conceivably, in the worldwide Internet, a data packet, though encoded, may find itself propagating through a node or a set of nodes in an insecure region of the world where it may be intercepted by a hostile unit. While there is always a finite probability, however small, that the hostile unit may successfully break the cryptographic technique, even if the coding is not compromised, the hostile unit may simply destroy the packet, thereby causing the end systems to trigger retransmissions, which, in effect, slows down the network and constitutes a performance attack. The philosophy underlying the security approach in the Internet may be traced to the end-to-end reasoning in the survey paper by Voydok and Kent [67]. They are cognizant of the need to protect the increasing quantity and value of the information being exchanged through the networks of computers, and they assume a network model in which the two ends of any data path terminate in secure areas, while the remainder may be subject to physical attack. Accordingly, cryptographic communications security, i.e., link encryption, will defeat wiretapping. Furthermore, to defeat intruders who are otherwise legitimate users of the network, authentication and access-control techniques are essential. Voydok and Kent state a crucial assumption: For successful link encryption, all intermediate nodes—packet switches and gateways—must be physically secure, and the hardware and software components must be certified to isolate the information on each packet of data traffic transported through the node. The difficulty with the assumption in today's rapidly expanding, worldwide,

Internet is clear. Increasingly, however, researchers are criticizing the overemphasis on cryptography and are stressing the need to focus on other, equally important, aspects of security, including denial of service and attacks aimed at performance degradation. Power [26] warns of a new kind of threat, information warfare, which consists in disabling or rendering useless the enemy's key networks including the command and control, power grid [20], financial, and telecommunications networks. It may be pointed out that the literature of the 1970s and 1980s contains a number of references to many of the noncryptographic security concerns that had been proposed primarily for operating systems. Thompson [68] warns of the danger of compiling malicious code, deliberately or accidentally, into an operating system and labels them Trojan horses. In enumerating the basic principles for information protection, Saltzer and Schroeder [69] warn against the unauthorized denial of use and cite, as examples, the crashing of a computer, the disruption of a scheduling algorithm, and the firing of a bullet into a computer. They also propose extending the use of encipherment and decipherment beyond their usual role in communications security to authenticate users. In stating that concealment is not security, Grampp and Morris [70] reflect the reality that computer systems ought to remain open, and clever techniques must be invented to ensure information security.

Security in ATM networks is a recent and rapidly evolving phenomenon. The current literature reveals two principal thrusts: the use of cryptography to encrypt ATM cells and the use of digital signatures in authenticating end-to-end signaling. In June 1996, Cylink and GTE [71] were the first to demonstrate a cell encryptor for ATM networks. Spanos and Maples [72] have proposed an MPEG video compression algorithm to achieve security of multimedia traffic in ATM networks. Chuang [73] proposes access control for secure multicast in ATM networks. Wilcox [74] describes the experiences with the ATM testbed ATDNet and reports that the four coupled areas of interest include ATM interoperability, distributed computing, information security, and high-speed network connections. Deng, Gong, and Lazar [75] propose mutual end-to-end authentication in signaling, cryptographic key distributions, and data protection toward security in ATM networks. Stevenson, Hillery, and Byrd [76] propose the use of cryptography to achieve data privacy and digital signatures to authenticate the end users during the setup procedure. They claim that a secure call setup must complete within 4 seconds of initiating a request. The encryption of ATM cells and the authentication in signaling, however, are only two elements of the much bigger picture of ATM network security.

In their presentation of the ATM Forum's approach to network security in ATM networks, Peyravian and Tarman [77] propose the use of authentication, key exchange, and negotiation of security options in an end-to-end manner as in the case of the Internet. Thus, their proposal inherits the basic difficulties of the peer-level, end-to-end approach as explained earlier in this section. Furthermore, the lack of confidence in the lower levels of the network OSI model forces Peyravian and Tarman to require, even after the virtual circuit is established and the propagation of traffic packets is initiated, that some form of message authentication code be appended to every ATM packet. This results in high overhead and loss of efficiency.

An additional serious difficulty is that the end-to-end network security approach may not be engaged until a virtual path has first been successfully determined by the call setup process. Peyravian and Tarman [77] acknowledge this requirement while discussing ATM security messaging. For there is always a chance that a call setup process may not succeed. However, once the virtual path has been determined, the intermediate ATM nodes have already been committed, and it may already be too late in that many of the potential security considerations may no longer be available. The key difficulty with the ATM Forum's proposal stems from the desire to adopt security techniques that have been developed for data networks that, given the unique characteristics of ATM networks, may not constitute a scientifically proven path.

The most recent literature [78][79][80][81] in network security focuses on the use of cryptography both to encrypt the data cells and for end-to-end authentication in the Internet, while Tarman, Hutchinson, Pierson, Scholander, and Witzke [82] review encryption algorithms for fast encryption of cells in ATM networks. The ATM Forum has yet to publish the complete specifications for network security. In summary, the current literature on ATM network security lacks a principle that can deliver comprehensive security. This chapter addresses the issue by coupling the fundamental security framework with the unique and fundamental characteristics of ATM networks.

3.2   The Need for User-Level Security on Demand

Fundamentally, under security on demand, a network allocates security resources to a user, commensurate with the user's request and resource availability, instead of assigning the same security resources to all users. As a result, a wide range of user requests may be supported, which may be, in general, efficient and, at times, critical. For instance, where the email greetings sent by a commander to the troops in one battlefront are sent on the same network that is carrying sensitive intelligence data and classified commands for a different battlefront, both messages will require identical and expensive security resources. Conceivably, the greetings may consume valuable security resources and deny service to the classified commands, jeopardizing the second front. It is likely that the users' needs will differ quickly and substantially, and therefore the security on demand paradigm must be dynamic. In a network, security resources are associated with the geographically dispersed nodes and links, which implies that the approach must undertake the allocation of the distributed resources dynamically. The allocation task is complicated by two facts. First, security resources are expensive and economic considerations may imply a nonuniform distribution throughout the network. Second, the users may interact with the network at different geographical points, asynchronously, i.e., at irregular intervals of time. This, in turn, requires that the allocation be based on a reasonable and scientific principle. Though not a requirement, logic dictates that a distributed allocation strategy is likely to be more efficient and less vulnerable than a centralized mechanism. By design, the telephony and ISDN networks allocate identical resources to all users, and as a result, these paradigms may be inappropriate for security on demand. A user message may be composed of one or more cells, the transport of each one of which must be subject to a uniform level of specified security. Thus, the Internet paradigm wherein different cells of the same message may be routed through different paths, not known a priori, poses a serious difficulty to the security on demand approach. In today's rapidly expanding Internet, security resources may be completely missing at many nodes and links.

A key advantage of user-level security on demand is that it enables the successful and secure transport of multiple-user traffic, each with its unique security requirements, simultaneously over the same network, down to the user-level. This, in turn, eliminates the need to provide the highest security to every network node: a prohibitively expensive proposition. Also, in a network containing both kinds of resources, users who do not need security may be provided with the nonsecure resources, thereby relieving the burden on the generally scarce security resources. Thus, the approach offers the perspective that security is a distributed network resource that is allocated to each user call based on demand and dictated by need.

3.3   Security on Demand in ATM Networks

3.3.1   Unique Characteristics of ATM Networks

The primary motivations underlying the design of ATM networks are three in number, and they translate directly into three unique characteristics. First, all traffic—audio, video, and computer data—are organized into fixed-size packets, termed ATM packets. Second, to deliver the level of performance requested by a user, the network will first examine the state of its available resources, then determine a qualified route and channel where possible, and finally initiate the transfer of the user traffic over the selected route and channel. Third, although technically the channels over a link between two specific nodes are allocated to users' traffic, the network will attempt to maximize the efficiency of resource usage through statistical multiplexing, i.e., carrying a user's traffic in excess of the agreement through a different channel that is less than fully utilized. The thinking is that where traffic is bursty and accurate prediction of the required resources is difficult, the network will attempt to dynamically optimize the resources.

From the perspective of this chapter, the second characteristic of ATM constitutes the key to enabling security on demand. Clearly, there is a distinct separation between the route and channel setup phase and the traffic transport phase. Although the first phase is termed signaling, analogous to telephony, and includes other complex functions, a key component is call setup. The basic principles involved in setting up a call were first introduced by Sato, Ohta, and Tokizawa [83], elaborated by Hac and Mutlu [84], and implemented by Chai and Ghosh [85] for representative ATM networks. Most recently, the ATM Forum has organized the specifications into the private network–network interface (P-NNI) 1.0 [86]. While the PNNI defines most of the elements of call setup, some key research issues have been left unresolved, many of which will be addressed subsequently in this chapter. Sato, Ohta, and Tokizawa [83] define a virtual path as a labeled path, i.e., bundles of multiplexed circuits, between virtual path terminators. Here, users are assumed to play the role of the terminators, and thus virtual paths are labeled end-to-end paths. When a user requests a call setup, the source node that intercepts the call will either identify an already existing path or will establish one, dynamically, based on demand and the user's specific requests. The route establishment is a local, computational process where the source node executes Dijkstra's shortest-path algorithm [87], utilizing its current knowledge of the links' characteristics, to obtain a route from the source to the destination node that optimizes a given criterion. In the simple scenario, the criterion may refer to selecting a route where the cumulative physical propagation delays of the constituent links is a minimum. This chapter proposes a more complex criterion which is discussed subsequently. Since the necessary resources are geographically dispersed and since their state is continuously evolving, the source node will actually transmit a setup message that includes a designated transit list (DTL) packet, all the way to the other end, both to ensure that the resources are actually available along the path and to reserve them for use by the traffic subsequently. The user traffic is launched when the test packet returns successfully. Otherwise, the call fails. Thus, the deliberate choice of the intermediate nodes and the control exercised over the transport of the traffic cells through these nodes is unique to ATM and provides immunity to ATM networks from the difficulties that are associated with data networks.

3.3.2 Integrating the Unique Characteristics of ATM Networks with User-Level Security on Demand

The fundamental contributions of this chapter are twofold. First, building on the fundamental security framework, it unifies the security concerns among all users and network elements through a common language. While the comprehensive security status for every node and its associated links is expressed through a 72-element matrix, every user is permitted to specify the security requirements for its message, comprehensively, through a 72-element matrix. Thus, security may be viewed as a new quality of service (QoS) metric in addition to bandwidth, end-to-end cell delay, jitter, cell loss, etc. Second, during the call setup process for a user, termed call admission control (CAC) by the ATM Forum, the user-defined matrix is examined against those of the appropriate nodes to guide the choice of the virtual path, and to ensure the availability of the security resources as well as reserve them. Thus, the element of security is integrated into the ATM call setup process, and the approach enables the network to provide personalized security on demand service dynamically to every user.

The details of the conventional call setup process, as specified by the ATM Forum, may be described as follows. When a user launches a call, the source node that first intercepts the call assumes the responsibility to obtain a route to the destination node, where possible. For geographic proximity, geopolitical, and other reasons, the nodes of an ATM network may be organized into groups where the intra-group connectivity information is held strictly local to every group. However, updated information on inter-group connectivity is maintained within every group, thereby enabling any node within a group to compute a path, where available, in terms of other group identifiers to the destination group that contains the destination node. Clearly, for a source node, the path within its group may be computed in terms of the nodes of the group, while the detailed paths within the intermediate groups are computed locally by the corresponding groups. Thus, the route computation may be viewed, conceptually, as a two-step process. First, the source node computes the overall route across the groups, starting from its own group to the group where the destination node is located. Second, the source node computes the route through its own group. The determination of the exact route across each of the intermediate and destination groups is the responsibility of the corresponding groups and occurs later, as explained subsequently. As stated earlier, the ATM Forum specifies a skeleton protocol, PNNI [86], to characterize the route determination process. Under PNNI, the choice of a function with the arguments—physical propagation delay of the ATM links, processing delays of the ATM nodes, etc., in the determination of a route, is left open. Once the route is computed successfully by the source node, the ATM networking model requires the source node to launch an actual setup message that includes the designated transit list (DTL) and one that is required to traverse through each of the actual ATM nodes in the appropriate groups, reserving the corresponding QoS resources, where available. In the event that the setup message fails to reserve the user required bandwidth, the message returns as unsuccessful and the call fails. Where the setup message succeeds in reserving the bandwidth all along the virtual path, up to the destination node, the call processing is viewed a success and the user traffic is then launched on the corresponding switching network.

The call setup process, augmented with security on demand, is described as follows. To quickly recapitulate, first, every ATM node and its links are character- ized by a matrix. Second, every user call request is accompanied by a matrix, the content of which represents the user's desired security characteristics for the call. Third, under CAC, the user-specified matrix and the matrices of the nodes play a predominant role in determining the source node's computation of the virtual path. The exact function is left up to the implementation, and its argument(s) may include either one or more of the following: security resources at each node and link, the bandwidth available on a link at a given time instant, processing delays at the nodes. In addition, the call setup message, which includes the DTL packet, carries with it the user-specified matrix and compares it against the se- curity matrices of the nodes along the virtual path. Should every ATM node and link along the computed virtual path meet the user-specified security requirement and have available bandwidth, a secure path is successfully determined from the source to the destination node. ATM traffic cells may be transported over the cor- responding switching network. Otherwise, where any of the intermediate nodes, including the source and destination nodes, fail to meet the user-specified security requirement, the call fails. Following the completion of transport of the ATM cells through the network, the network resources are released for use by future call requests. To summarize, the security on demand approach realizes the dynamic creation and tear-down of routes at different security levels and the simultaneous existence of multiple routes in the network, each rated at a different security level.

As an example of a secure call setup under security on demand, consider the 9-node representative military ATM network shown in Figure 3.1. In Figure 3.1, while the White House ATM node is connected solely through the Pentagon node, the latter is directly connected to the downtown D.C. node and the Alexandria node, both of which are directly linked to Norfolk. Other military nodes are located at Baltimore, Ft. Meade, Andrews Air Force Base, and the Naval Academy. Assume that for every link, its security resource is encapsulated in detail through a 72-element matrix, as shown in Figure 3.1. Each element of a matrix may assume a value between 0 and 9, with 0 expressing the most stringent security available and 9 implying the total absence of any security. In addition, for each link, the largest value of any of its 72 matrix elements serves as the composite, top-level security index, which in turn, is known by every node and utilized during the augmented call setup process. The composite security index reflects a conservative view and identifies the element with the weakest security. Consider a scenario where the President, located at the White House node, intends to send a Top Secret message to the Commander in Chief (CinC) of Atlantic Command, located at Norfolk. Clearly, the route selected for this Top Secret message, the only one possible in this network, will include the intermediate nodes Pentagon and Alexandria, since the corresponding links offer composite security index values of 0. The route is highlighted in Figure 3.2.

Figure 3.3 presents a representative joint military and civilian ATM network that bears the same topology as that in Figure 3.1 but where only a few links are rated at 0, i.e., most secure. The remainder of the links are labeled 9, completely

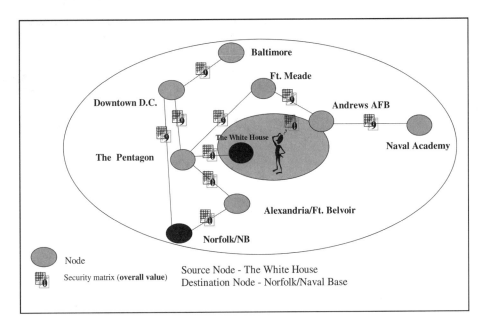

Figure 3.1 Illustrating a secure ATM call setup between the White House and Norfolk in a representative military network.

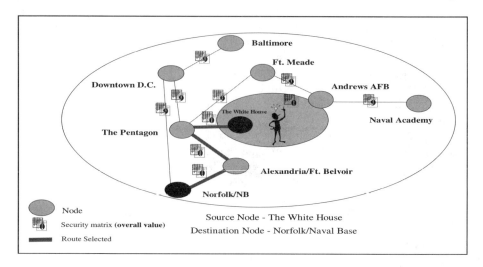

Figure 3.2 A secure ATM call established between the White House and Norfolk in a representative military network.

insecure. Consider two military users A and B, both located at the White House and interested in transporting traffic to the Norfolk Naval Base. While user A insists on a secure route from the White House to the Norfolk Naval Base, user

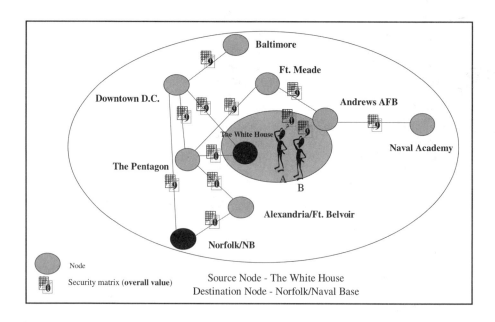

Figure 3.3 Illustrating secure ATM call setup between the White House and Norfolk in a representative joint military and civilian network.

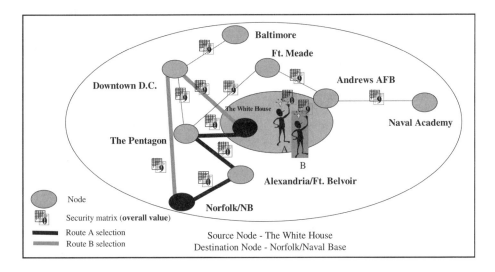

Figure 3.4 Secure and insecure ATM call establishment between the White House and Norfolk in a representative joint military and civilian network.

B needs no security. As in the case of Figure 3.2, the secure traffic for user A propagates through the Pentagon and Alexandria intermediate nodes, and the route is shown in Figure 3.4, as a darkened black line. The unclassified DoD traffic is transported through the downtown D.C. node, and is also shown in Figure 3.4 in gray.

To assess its practicality and usefulness in the real world, the proposed security on demand approach is subject to rigorous scientific analysis through modeling, asynchronous distributed simulation, and performance analysis. This is the subject matter of the next chapter.

3.4 Problems and Exercises

1. Given that the fundamental security framework constitutes a scientific and logical approach to measure a network's security as well as permit an individual to express the security needs, can it be integrated into an IP network. Recall that an IP network is fundamentally a store-and-forward type of network? Explain your reasons.

2. Repeat the previous question for the classic telephone networks and contemporary ISDN networks.

3. Discuss the advantages and limitations of the notion of security on demand.

4. Suggest substantial refinements to the security on demand approach.

5. Synthesize scenarios where neither the static certification-based security nor the dynamic security on demand, approach is appropriate. Suggest an alternative approach under these circumstances.

4

The Concept of Node Status Indicator (NSI) and Performance Analysis

4.1 Introduction

Traditionally, incorporating security into a system has been an afterthought, and the techniques employed have generally been ad hoc. As a result, performance degradation is frequently observed to accompany secure systems. In the network security literature, Runge [88], Hale [89], and Stacey [90] observe that while the current solutions provide security in specific areas, they give rise to performance degradation, fail to allocate resources optimally, and are expensive. A more serious difficulty has been the lack of detailed reports in the network security literature describing the validation of the proposed security mechanisms prior to developing an actual prototype and their performance analysis, especially for large-scale representative systems. Stevenson, Hillery, and Byrd [76] propose the use of cryptography to achieve data privacy and digital signatures to authenticate end users during the setup procedure. Their claim that a secure call setup must complete within 4 seconds of initiating the request is neither accompanied by any scientific validation nor related to any published scientific standard. In principle, the 4-second mandate, as stated, is difficult to generalize, since the call setup time is a function of the relative locations of the source and destination nodes, i.e., whether they are intra-group, inter-group, etc.

This chapter focuses on the issue of validating proposed security mechanisms for ATM networks through (1) accurate behavior modeling of a representative ATM network, (2) executing a large-scale, asynchronous, distributed simulation of the model on a loosely coupled parallel processing testbed with stochastically generated synthetic input traffic, and (3) analyzing the simulation results through innovative performance metrics. Given that the nodes of a real ATM network interact with each other and the users asynchronously, i.e., irregularly in time, the execution of the asynchronous distributed simulation approach on the testbed resembles an operational system. Consequently, the performance analysis can yield valuable insights into the complex network behavior.

4.2 The Concept of Node Status Indicator (NSI): Refining Distributed Resource Allocation

In an ATM network, the resources include the nodes that provide the computational and decision-making ability to process secure call requests and the links that provide the secure bandwidth. For calls originating at a node and destined for a distant node, the utilization of both local resources and those far away may be necessary. Since users interact with the total network dynamically, asynchronously, and at different geographical points, the resource availability scene is likely to be highly dynamic. Ideally, for efficient resource allocation every node must possess knowledge, at every time instant, of the exact state of the network, i.c., of all nodes in the system. The principal components of the state of a network include the network topology, i.e., the location and connectivity of the nodes and links, the physical propagation delays of the links, the security posture of every node and link, and the available bandwidth at the links. Conceivably, changes to the topology occur very slowly, relative to the rate at which call requests are asserted and processed by the network. Therefore, it is logical to assume that every node maintains a relatively accurate picture of the network topology. This, in turn, implies that to determine a route, one may take into consideration only the physical link propagation delays, which is static information. This approach is implied by Sato, Ohta, and Tokizawa [83] and is logical, since in ATM the call setup packet including the DTL is launched by the source node in any case to check for resource availability and reserve the necessary resources. Also, the nodes' and links' security statuses, encapsulated through the matrices, are essentially static information. Therefore, this information, once propagated across the network during initialization, may be viewed similarly to the physical link propagation delays.

However, the bandwidth availability associated with the links may experience rapid fluctuation, and this chapter proposes to utilize updated bandwidth availability information to guide the route determination process. The resulting refinement to route determination constitutes a sound, scientific strategy, and is likely to improve efficiency and reduce the probability of call setup failure. This research proposes the propagation of the available bandwidth at its links, encapsulated through a node status indicator (NSI) value, by every node to all other nodes, through the periodic use of the flooding algorithm [91]. The information is propagated through special communications links, termed signaling links. In a sense, the NSI encapsulates the state of a node, and its definition and rationale will be described subsequently. Under flooding, a node computes and propagates the NSI value to all of its neighbors, each of which then forwards the information to its neighbors, and so on until, eventually, all of its peer nodes in the system receive the information. Clearly, the exact times at which the nodes receive the information will differ. This time interval between the receipt of information at a node and the propagation from the original node is referred to as data latency, and its origin lies in the fundamental laws of physics. Latency causes uncertainty and is one potential cause of a call setup message to fail. Also, the farther a node is located away from the original node, the later it will receive the node's information, which will contribute to a loss in accuracy, in the sense of timeliness. While a high rate of flooding may enable a node to possess a more accurate picture of the state of the overall network, it may also cause congestion and overwhelm the computing engines at the nodes. In contrast, a very low rate of flooding may not impart meaningfully accurate knowledge of the state of the network, especially where it is highly dynamic. The choice of the flooding interval utilized in this study and the underlying rationale are described later in this chapter. It is assumed that an ATM network is organized as a two-level hierarchy and realizes the following flooding strategy. Within each group, every node disseminates its information to its peers, i.e., within its own group, through flooding. Clearly, this exchange of information involves only the intra-group links. Flooding is not realized at the next higher level, i.e., the peer group leaders do not disseminate information on the inter-group links to each other. The reason is as follows. During computation of the route across the groups from the source to the destination nodes, the issue of reducing the hop count, i.e., the number of intervening groups, is viewed as more important than utilizing updated information through flooding, especially if the latter leads to an increase in the hop count. The strategy is simpler, and the associated computational demand is low.

Every node computes an NSI value for each of its links, at intervals equal to the flooding interval, and propagates it throughout its group. As explained earlier, NSI values are computed only for the intra-group links. In addition, every node stores the information received from other nodes within its group, relative to the NSI values for their corresponding links. The information is organized within every node through an NSI table and used when the node attempts to obtain a route to a destination node by executing the Dijkstra shortest-path algorithm. The table is updated at flooding intervals and is dynamic.

The rationale underlying the definition of NSI is as follows. The state of a network consists of the following components: the physical propagation delays of the links, node processing delay for all nodes, queuing delay in the buffers at the nodes, security posture of the links, and the available bandwidth at the links. The dissemination of an excessive number of information items will congest the signaling links and impose significant computational demand at the nodes. It is noted that both the processing delay at a node and the queuing delay of the buffers at the node will be influenced directly by the traffic and thus may be accounted for by the bandwidths consumed at the links. The key components of NSI include the physical propagation delays of the links, security posture of the links, and the available bandwidth at the links. This research argues that in determining a secure route, the influence of the physical link propagation delays should be lessened. Otherwise, those route(s) that bear the lowest cumulative link delay from the source to the destination node will be selected more often, for different call requests, leading to an unbalanced use of the overall network resources. The physical link propagation delays should not be eliminated completely from consideration, since they will invariably have an impact on the end-to-end cell delay performance. Both the security posture and the available bandwidth of the links are equally important parameters in determining a secure route. Together, they will guide the Dijkstra shortest-path algorithm to explore a greater number of alternative routes through the network. For every link, the highest value of all the 72 elements is used as a conservative, i.e., most vulnerable, representation of the link's security posture. This value ranges from 0 to 9. The available link bandwidths for links are normalized according to the following step function:

Normalized available link bandwidth =

$$
\begin{cases}
9, \text{ when consumed bandwidth} < 9.68 \text{ Mb/s,} \\
6, \text{ when } 9.68 < \text{consumed bandwidth} < 19.375 \text{ Mb/s,} \\
3, \text{ when } 19.375 < \text{consumed bandwidth} < 38.75 \text{ Mb/s,} \\
1, \text{ when } 38.75 < \text{consumed bandwidth} < 77.5 \text{ Mb/s,} \\
0, \text{ when } 77.5 < \text{consumed bandwidth} < 155.5 \text{ Mb/s.}
\end{cases}
$$

The normalized physical intra-group link propagation delay = propagation delay of the intra-group link (in ms) mod 100.

Given that the highest physical propagation delay of any intra-group link in any group in the representative ATM network used in this study is 488 microseconds (μs), the normalized physical link propagation delay will range from 0 to 4.88, which is approximately one-half the range of the values for the security posture and normalized available link bandwidth parameters. Therefore, the NSI is computed as follows:

NSI, for a link = security posture of the link + normalized available link bandwidth + normalized physical link propagation delay.

As an alternative, one could propagate the static values of the security posture and the physical propagation delays of the links throughout the network at initialization. Only the dynamic information—available link bandwidth—may be flooded periodically throughout the network. Upon receipt, a node may then compute the current NSI values for all the links and use them during the Dijkstra shortest-path computation. A limitation of this approach is as follows. While there may not be a significant reduction in the size of the information disseminated through flooding, the increase in the computational demand at the nodes may be significant.

4.3   Modeling Security on Demand for a Representative ATM Network and Implementation Issues

The security on demand approach is modeled for a 40-node representative military ATM network, shown in Figure 4.1. The network topology, i.e., the locations of the nodes and the links that interconnect them, is synthesized utilizing unclassified knowledge of the actual locations of major defense installations and processing centers. It is representative of a military, continental US (ConUS) ATM network

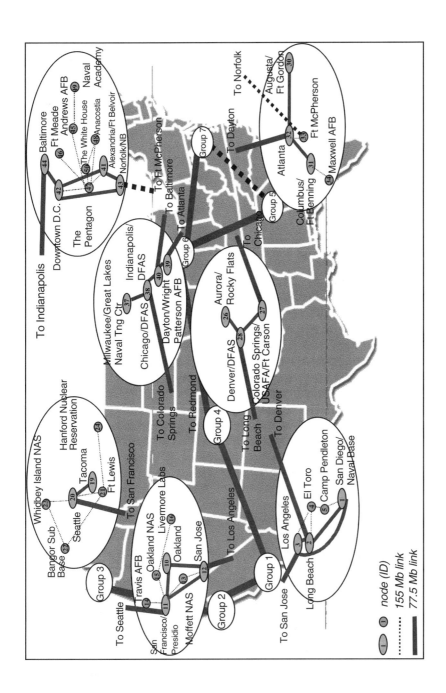

Figure 4.1 A continental US (ConUS) 40-Node representative military ATM network.

for carrying Top Secret traffic. Thus, the security posture of every link is set at the highest value, namely 0. The 40 ATM nodes are organized into a total of seven peer groups, based on geographic proximity. The inter- and intra-group link capacities are rated at either 155.5 Mb/s, or 77.5 Mb/s, and the assignments are partially stochastic. The physical propagation delays of the links are computed (in ms) through dividing the straight-line distances between the nodes (in km) by 194,865 km/sec, the speed of light in fiber [92]. Figure 4.1 presents the link propagation delay values and the bandwidths for all of the links.

In the model, the behavior of each ATM call processing node is encapsulated through a Linux process, the links between the ATM nodes are represented through TCP-IP connections between the corresponding processes, and the transport of the call setup messages is modeled through guaranteed messages between the processes. In essence, the key elements of the ATM Forum's specifications—PNNI 1.0 and UNI 3.0 for the nodes—are encapsulated in the node behaviors. The transport of ATM traffic cells is not modeled here, and consequently, the corresponding components in PNNI 1.0 and UNI 3.0 are absent. In an ATM network, the operation of the nodes is not synchronized, and the users interact with the system in an uncoordinated, asynchronous manner. To ensure the accurate execution of the asynchronous events, the processes are executed through a conservative, null message based, asynchronous distributed simulation algorithm [93]. The value of the timestep, a fundamental quantity in the simulation, is set at 2.74 μs, and it reflects the fastest subprocess in the network, the 155.5 Mb/s link. The choice of the timestep guarantees that every subprocess in the network may be accurately represented in the simulation. In summary, the accuracy of the modeling is ensured by the accuracy of the behavioral model of the ATM call-processing node, the simulation-algorithm-guaranteed correct order of execution of events, and the choice of the timestep.

The simulation is executed on a testbed consisting of a network of 35+ 90-Mhz Pentium workstations under the Linux operating system, configured as a loosely coupled parallel processor. The reasons for using the testbed are as follows. First, the clocks that control the execution of the processors are not coordinated. As a result, the Linux processes residing in the processors are executed asynchronously, while messages are exchanged between the processes via the Fast Ethernet, also asynchronously. Thus, the simulation is unique in that its execution closely resembles an operational ATM network, which, as explained earlier, is also asynchronous by design. Second, for a systematic study of the performance impact of security on demand, it is essential to accomplish hundreds of large-scale simulation runs within a realistic time frame, with each simulation run corresponding to different choices of network parameters and input traffic. The testbed executes the simulations quickly and permits this objective to be achieved.

The ATMSIM 1.0 [94] simulator is written in C/C++ and consists of over 15,000 lines of code. Each simulation run executes for 1.2 million timesteps (3.28 seconds), requiring approximately 58 hours of wall clock time for completion. That is, the longest-running processor finishes within 58 hours. Each 90-Mhz/200-Mhz Pentium workstation is equipped with 64 Mb of RAM and 1.2 Gb of disk space. ATMSIM 1.0 accurately models every important characteristic of ATM Forum's PNNI 1.0 (private network–network interface) and UNI 3.0 (user network interface).

4.4 Synthesis of the Input Traffic for Network Performance Analysis Under Security on Demand

The choice of the input-traffic and other network input parameters in network simulation is crucial to obtaining realistic performance data and useful insights into the network behavior. Traffic generation represents a careful tradeoff between the goal of exposing the network to worst-case stress and examining its behavior and the need to ensure a stable operational network, one that executes continuously, 24 hours a day, 365 days a year, without failing. While much work has been reported in ATM traffic modeling, the literature on call setup requests is sparse. Given the lack of large-scale operational ATM networks in the public domain, operating under the mode of switched virtual circuits, actual data on call setup requests from operational networks are difficult to obtain. In this research, while the synthetic traffic is stochastic, it is designed to resemble an operational system. The key input traffic parameters include (1) call arrival distribution, (2) bandwidth distribution in the calls, (3) call duration distributions in the calls, (4) security distributions in the calls, and (5) traffic mix, i.e., the relative percentage of inter- and intra-group calls in the entire network. With the exception of item 5, the distributions in items 1 through 4 are generated stochastically, and the choice of the key parameters is explained subsequently. Following their generation, traffic stimuli are saved in files for use during simulation.

4.4.1 Call Arrival Distribution and Network Stability Criterion

The call arrival distribution will bear a direct relationship to the stress level of the network. The higher the number of calls within the given length of simulation time, the greater the demand on the network resources, leading to stress. Excessive stress may be manifest through two key behaviors. First, the call setup time between any given source and destination node pair may increase consistently for successive calls, i.e., as time progresses, that are inserted into the network. This implies an unstable network, since as the operation progresses, the measure of the call setup time will ultimately exceed any prescribed threshold and be deemed practically unacceptable by the user. Such high stress is likely to lead to network failure. Second, the call success rate for the total network may decrease beyond a measure that the network provider finds unacceptable.

This research presents a novel two-part strategy to determine an appropriate call arrival distribution for the network modeled. In the first part, under this strategy calls are generated through a Poisson distribution function, and the network is simulated for different choices of the mean value of the distribution. For each simulation experiment, graphs of the call setup times as a function of simulation time, for every pair of source and destination nodes, are obtained and analyzed. Where any of the graphs exhibit a nonuniform behavior with the call setup time increasing consistently with the progress of simulation, the network is considered to be driven into an unstable region by the excessive call arrival density. The argument is that for networks to remain operationally stable, a graph of the call setup time as a function of simulation time, for any given pair of source and destination nodes, must remain uniform. Where all of the graphs exhibit uniform behavior, the network is considered to be within the stable operating region. Through trial and error, i.e., by executing a number of simulations for different choices of the mean of the Poisson distribution, this study yields a call arrival distribution that stresses the network to the edge of stability.

Under the second part, the call success rate for the call arrival distribution obtained under part 1 is measured and examined. Call success rates in the neighborhood of 90% are considered to reflect acceptable operating conditions, and the corresponding call arrival distribution is utilized for performance analysis.

Analysis of preliminary simulation results reveals that the node pair (3, 2) in group 1 incurs high values for the call setup time, and is selected as the focus for the stability study. Since a significant bulk of the calls are local, it is not surprising to find the stability study focused on an intra-group node pair. In Figure 4.1, while group 4 has very few links, implying poor connectivity, group 7 reveals high connectivity among the nodes. The nodes of group 1 are connected through a reasonable number of links, and this group is selected as representative of typical groups in a military network.

Figure 4.2 presents the topology of group 1 of the 40-node representative ATM network in Figure 4.1. The inter-group links are not shown here, since they do not constitute the focus of the stability study.

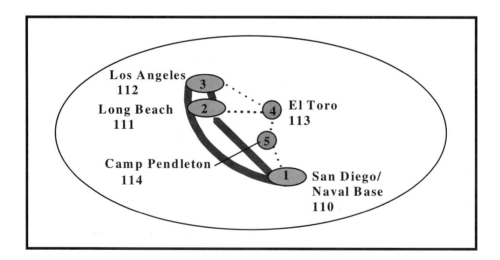

Figure 4.2 Determining acceptable call interarrival interval through stability study.

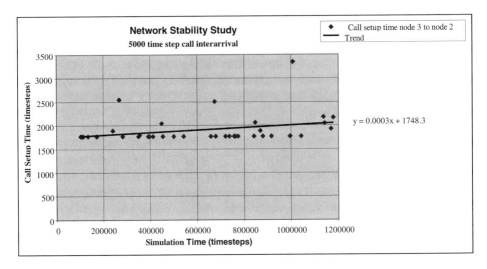

Figure 4.3 Call setup times for calls between nodes 3 and 2, as a function of the simulation time (in timesteps), for interarrival interval = 5,000 timesteps (13.7 ms).

Figures 4.3 through 4.6 present plots of the call setup time values between nodes 3 and 2, as simulation progresses, for interarrival intervals of 5,000 timesteps, 7,500 timesteps, 10,000 timesteps, and 15,000 timesteps, respectively. User call

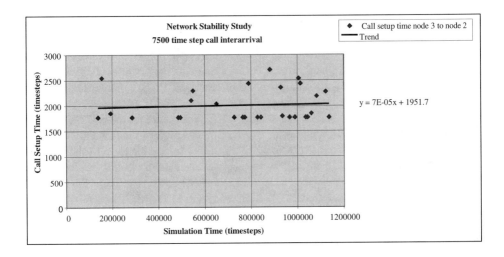

Figure 4.4 Call setup times for calls between nodes 3 and 2, as a function of the simulation time (in timesteps), for interarrival Interval = 7,500 timesteps (20.5 ms).

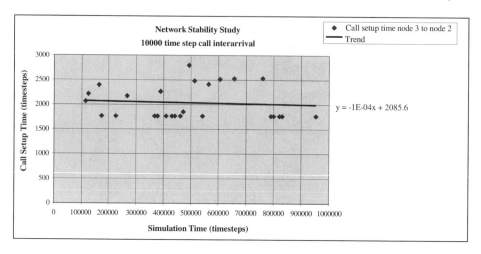

Figure 4.5 Call setup times for calls between nodes 3 and 2, as a function of the simulation time (in timesteps), for interarrival Interval = 10,000 timesteps (27.4 ms).

requests are inserted into the network once initialization is complete, at 90,000 timesteps, and this continues up to 1,200,000 timesteps, when the simulation is terminated. For each of the plots in Figures 4.3 through 4.6, the straight line that best fits the data is computed and shown on the plots. Observations indicate

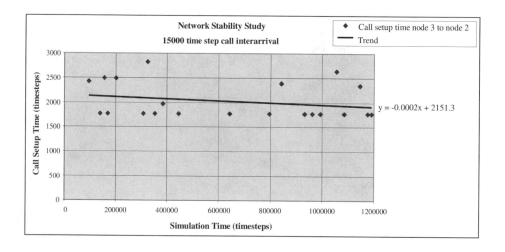

Figure 4.6 Call setup times for calls between nodes 3 and 2, as a function of the simulation time (in timesteps), for interarrival Interval = 15,000 timesteps (41.0 ms).

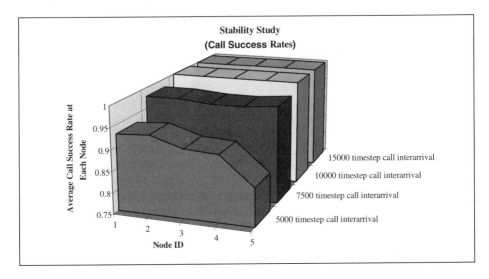

Figure 4.7 Call success rates as a function of the nodes, for call interarrival intervals of 5,000, 7,500, 10,000, and 15,000 timesteps.

that the slope of the best-fit line in Figure 4.3 is positive, implying an unstable network. The slope in Figure 4.4 is nearly 0, while those in Figures 4.5 and 4.6 are slightly negative, reflecting stable operating conditions. Clearly, the choice of 7,500 timesteps (20.5 ms) for the interarrival interval marks a significantly stressed though stable network.

Figure 4.7 presents the call success rates for all intra-group calls at each of the five nodes of group 1, for different choices of the inter-arrival intervals, ranging from 5,000, to 7,500, 10,000, and 15,000 timesteps. The corresponding call success rate values are approximately 90%, 98%, 100%, and 100%. From the results in Figures 4.3 through 4.6 and Figure 4.7, the value 7,500 timesteps emerges as the choice for the interarrival interval.

4.4.2   Call Bandwidth and Call Duration Distributions

The bandwidths and durations for the synthetic calls used in this study are obtained through uniform distribution functions. The rationale underlying the choice of the parameters is expressed as follows. While each user is allocated one-eighth of the 64kb/s link capacity or 8 Kb/s of bandwidth in telephony, ISDN provides two 64 kb/s traffic channels to each user plus a 16 kb/s for signaling. The Internet principally utilizes combinations of T1 connections, each rated at 1.544 Mb/s, for its node-to-node links. Thus, a lower bound that may be used for call bandwidth choices is limited to 1.544 Mb/s. The 60–80 Mb/s bandwidth required for HDTV constitutes the upper bound for the choice of call bandwidths. This study assumes call bandwidths to range between 1 Mb/s and 10 Mb/s, with the average value at 5 Mb/s, and argues that in the future, link bandwidths are likely to increase quickly, implying a high probability that users will be offered greater bandwidths. Consequently, session durations will tend to decrease. Conceivably, the typical file for FTP and email may also incur an increase in size. Kleinrock [95] assumes a typical file for FTP or email of size 1 Mbit. Assume that in the future this parameter increases to 5 Mbits. The transmission duration at the average bandwidth of 5 Mb/s will approximately equal 1000 ms. This research assumes a uniform distribution for call durations with a range of 300 ms to 900 ms and average at 600 ms. The reason for choosing lower call duration values is to permit the network to accommodate a larger number of calls and operate under significant stress, a scenario that is of interest to this study. Although the author believes, as stated earlier, that future networks will offer much higher bandwidths to the users, implying a commensurate decrease in call duration, a total of 5% of all call requests in the network are characterized by indefinite duration, i.e., following initiation, they remain in effect throughout the remainder of the simulation. Such calls reflect a typical military need to keep open a few channels for critical communication.

4.4.3 Call Security Distributions

The literature on the security classification of user traffic within the military and government is sparse, and virtually nonexistent for commercial traffic. Within both the government and military, there are no definitive sources that quantify the volume of traffic that requires security. In the early 1970s, several congressional hearings were organized to focus on the issue of how much government information is classified and the methods used to determine the worthiness of protection. One of the more famous and widely reported hearings focused on the Pentagon Papers in 1973. Commenting on the hearings, Dorsen and Gillers [96] report, "[it is] clear that no one really knows just how many classified documents there are in any federal agency." Horton and Marchand [97] state that federal agencies generate the equivalent of 10 billion sheets of information a year. Dr. Rhoads, archivist of the US, testified before Congress that the US Archives contain 470 million pages of classified documents [96] accumulated over a 15-year period. This computes to an average of 31.3 million pages a year, or the equivalent of approximately 1.0% of all of the paper documents generated by all agencies in the US government. Horton and Marchand [97] also state that as a part of the federal budget, money spent on classified federal information industries ranged from 14.5% to 20.5% of the total information budget between 1961 to 1970. Based on the information reviewed here, this research assumes that the federal government generates classified information at a rate anywhere from 1% to 20% of the total of all information produced by the government. Building on this knowledge, this study assumes that 20% of the DoD traffic is classified and rated at Top Secret, or value 0. The remainder of the DoD traffic, namely 80%, is assumed to be unclassified or rated at value 9.

4.4.4 Traffic Mix: Inter- and Intra-Group Call Request Distributions

For the synthetic calls inserted at an originating node, termed source node, during simulation, the destination node identifiers are determined stochastically, with a cumulative probability of 75% that the target node is local, i.e., intra-group, and 25% probability that the destination is nonlocal, i.e., an inter-group node. For local call requests the distribution is uniform among the nodes of the source node's peer group. For nonlocal call requests, the distribution is uniform first among the other groups in the total network and, second, among the nodes of each group. The choice of the probability values reflects the argument that the groups must be organized primarily on the basis of the calling patterns. In the absence of data from an operational public ATM network, the calling pattern data contained in Mitchell and Donyo's report [98] "Utilization of the US Telephone Network" is examined. It reveals that while the number of long distance (toll) calls per capita is 15.35% of the total number of all calls made in 1991, the total number of minutes consumed by long distance calls is 30.35%.

4.5 The Design of Experiments, Simulation Results, and Performance Analysis

The principal objective of this section is threefold: (1) to demonstrate scientifically the feasibility of the concept of security on demand in ATM networks, (2) to analyze the performance impact of security on demand, and (3) to analyze the impact of NSI on the performance of the network. A number of simulation experiments have been designed and executed on the testbed to achieve goals (1) through (3).

For objective (1), the focus is on executing the 40-node representative military ATM network on the testbed and examining whether the secure call setup requests are established successfully. The corresponding simulation is referred to as the "baseline" version. It implements the key elements of ATM Forum's PNNI 1.0 and UNI 3.0 specifications. Thus, the Dijkstra shortest-path computation utilizes only the physical link propagation delays. It is pointed out that the PNNI specification does not mandate a specific path-selection algorithm. It states, "efficient QoS-sensitive path selection is still a research issue," and that multiple independent link parameters are "expensive computationally."

For objective (2), the corresponding simulation is referred to as the "security on demand" version. It also implements the key elements ATM Forum's PNNI 1.0 and UNI 3.0 specifications with the exception that it associates security matrices with links and nodes as well as user-specified calls.

For objective (3), the corresponding simulation is referred to as the "NSI augmented security on demand" version. It is identical to the "security on demand" version except for the following. First, each node disseminates the link bandwidth availability of its links through flooding. Second, the link bandwidth information is utilized by other nodes, along with the physical link propagation delays, to guide the Dijkstra shortest-path computation. The NSI information is flooded every 25,000 timesteps (68.5 ms). This interval is determined empirically, through extensive trial and error, and reflects the fastest rate of flooding just short of introducing significant congestion in the signaling links and imposing excessive computational demand at the nodes. At each node, a call processing delay of 800 timesteps (2.19 ms) is introduced to emulate realistic behavior of ATM nodes. This parameter is referred to as the call admission control (CAC) delay. Thus, every user call request will require 2.19 ms for processing at a source node. User call requests that are asserted at a node faster than 2.19 ms must be queued, and are processed in the order received. The value of the CAC delay is determined from three independent sources. First, based on the specifications reported in Network Computing [99], a Fore Systems, ASX 1000 ATM switch is observed to achieve a top routing speed of 437 calls per second, which is equivalent to approximately 1 call every 800 timesteps (2.19 ms). Second, instrumentation of the simulation reveals that the code segment emulating the call processing at a source node requires 11.9 ms on the 90-Mhz Pentium workstation. The equivalent computation time on a 450-Mhz Pentium processor would be reduced to 11.9/5=2.38 ms. Third, the state-of-the-art AT&T 5ESS telephone switch is rated at 1.2 million calls/hour (333 calls per second), which yields a call processing delay of 3.0 ms. Despite key differences between telephony and ATM, the processing speed of the 5ESS switch serves as a reasonably accurate predictor of the call processing delays in ATM networks.

For objectives (2) and (3), this research focuses on two metrics: call success rate and call setup time. While the call success rate measures the number of calls for which routes are successfully obtained through the network, the call setup time measures the total time, in timesteps, that the network requires to establish a route for a successful call. An additional measure, the average number of hops encountered by the successful calls in the network, is also utilized to assess the benefit of NSI.

4.5.1 Successful Integration of User-Level Security on Demand in ATM Networks

4.5.1.1 *Metric I: Call Success Rate.* Figure 4.8 presents the call success rates at each of the 40 nodes of the representative military ATM network, obtained through dividing the number of calls originating at the node that are successfully completed within the simulation duration by the total number of calls inserted at the node, and expressed as a percentage. The overall call success rate in the network is computed as 32.56%, based on 984 successful calls out of a total of 3022 calls. It is pointed out that the nodes in Figure 4.1 are not identified through contiguous integer values, as reflected in Figure 4.8. The call success rate values at the nodes take into consideration all calls at the nodes: intra- and inter-group. These values are significantly lower than those presented in Figure 4.7, and reflect the impact of the inter-group calls, many of which fail because of the presence of the "choke points," i.e., specific groups and nodes in the network through which many of the inter-group calls must pass. Examples include the nodes in groups

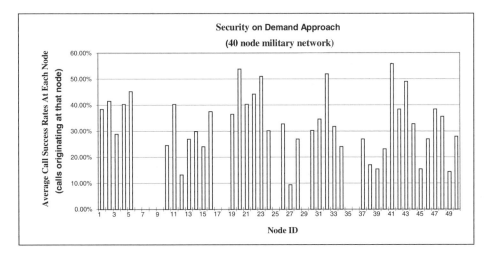

Figure 4.8 Call success rates as a function of the nodes of the ATM network, under security on demand.

4 and 6 in Figure 4.1. The choke points emerge as a result of poor inter-group connectivity in the network and arise from the military's goal of selecting an optimized network topology for economy and complementing it with alternative networks for greater overall reliability. Furthermore, of key importance in this research are the comparative call success rate values across the three approaches, not their absolute values.

4.5.1.2 *Metric II: Call Setup Time.* Figure 4.9 presents the mean call setup time, in timesteps, at each of the 40 nodes of the network, obtained through averaging the call setup times over all calls originating at a given node and successfully established within the simulation duration. Figure 4.9 reveals that the average call setup value ranges from a low of 2,610 timesteps (7.15 ms) at node 4 to a high of 8,167 timesteps (22.37 ms) at node 39. The average over all calls in the network is computed at 4,234 timesteps (11.60 ms). Analysis reveals that a significant number of the successful calls originating at node 4 imply short hop routes to other nodes within the group. This implies both a reduction in the computational delay for the calls and a lower cumulative physical propagation delay for the routes. Of a total of 42 successful calls at node 4, 30 require only one hop, while 11 require two hops to reach their destinations, and 1 call is destined for a different group.

At node 39, 1 of the 8 successful calls originating at this node requires a total of 11 hops to its destination, node 21, and a significantly high call processing delay of 35,748 timesteps. The statistics of this call plus the low number of successful calls at this node contributes to node 39 exhibiting a high average call setup time.

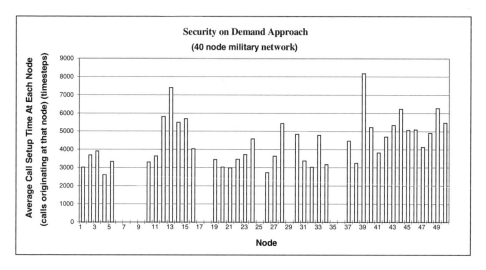

Figure 4.9 Average call setup times for calls originating at each node of the ATM network, under security on demand.

4.5.2   Comparative Analysis of the "Baseline" and "Security on Demand" Versions

For the baseline version, the resulting call success rates and call setup times for all of the nodes in the network are presented in Figures 4.10 and 4.11, respectively. While the overall call success rate is computed as 31.5%, with a total of 952 successful calls, the call setup time, averaged over all nodes and all calls, yields 4,246 timesteps.

A comparative analysis reveals that the call success rate for the security on demand is slightly higher, by 1.06%, relative to the baseline version and that the call setup time is slightly lower, by 12.0 timesteps, than the baseline version. Thus, the results corroborate the hypothesis that the performance impact of the user-level security on demand approach integrated into an ATM network is minimal.

Further analysis of the results under security on demand reveal that the high call success rates for 9 of the 10 nodes in group 7, labeled 41 through 50, contribute to the slightly higher performance of security on demand over the baseline version. Without the data from the nodes of group 7, the call setup rates for the two versions are nearly identical: 33.57% versus 33.21%, as would be expected. The differential performance of the nodes in group 7 for the baseline and security on demand approaches may be explained as follows. In the baseline approach, the Dijkstra shortest-path algorithm causes a route to be selected between a node pair that reflects the least cumulative physical link propagation delay. In contrast, under security on demand, the route determination is guided solely by the security posture of the links. Given that all links are rated highest, at 0, the computed routes are characterized by the shortest number of hops but not necessarily the

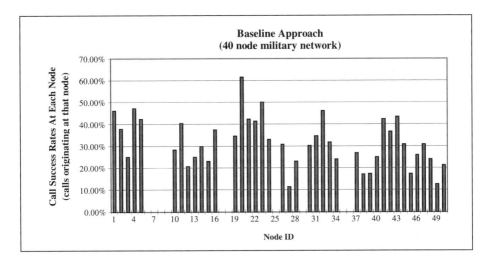

Figure 4.10 Call success rates as a function of the nodes of the ATM network, in the baseline version.

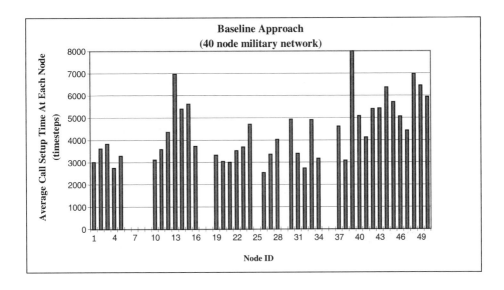

Figure 4.11 Average call setup times for calls originating at each node of the ATM network, in the baseline version.

least cumulative physical link propagation delay. For the given topology of group 7 and the distribution of the physical link propagation delays, security on demand appears to be favored, though marginally.

As an example, consider the progress of the call setup process for four successful calls to node 48 under the baseline and security on demand versions. For the baseline version, the four calls utilize the two routes shown in Figure 4.12. Three calls, with identifiers (say) 1, 29, and 39, select the 3-hop route from Alexandria through the Pentagon and White House nodes, terminating at Anacostia. The fourth call, say with identifier 73, selects the 2-hop route Alexandria through the Pentagon node to Anacostia. The route choices are influenced by the need to minimize the cumulative physical link propagation delay. In contrast, under security on demand, the goal of minimizing the number of hops causes a single 2-hop route to be selected for all four calls, as shown in Figure 4.13. Thus, the simulation results reveal that a route with fewer hops is more likely to succeed, since it will compete for bandwidth at fewer links and will incur a smaller contribution from the processing delays of the nodes. Furthermore, the key reason that group 7 amplifies the differences between the two approaches is to be found in its rich interconnection topology, relative to groups 4, 5, and 6, which provides multiple alternative routes between nodes.

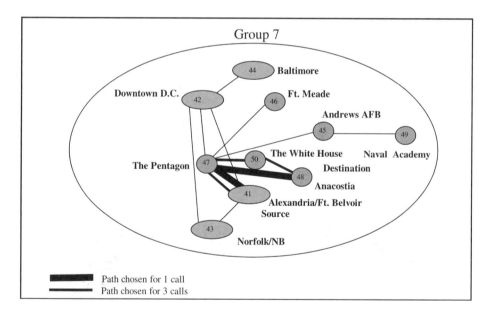

Figure 4.12 Route determination for four successful calls from Alexandria to Anacostia in the baseline version.

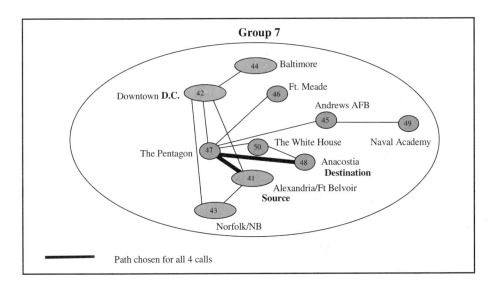

Figure 4.13 Route determination for four successful calls from Alexandria to Anacostia, under security on demand.

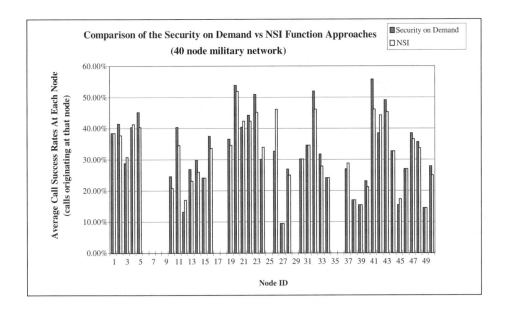

Figure 4.14 Call success rates as a function of the nodes of the ATM network, for both security on demand and NSI augmented security on demand approaches.

4.5.3   Analysis of NSI's Influence on Network Performance

Figure 4.14 presents the call success rates as a function of the nodes of the ATM network for the NSI augmented security on demand. The plot for the security on demand approach from Figure 4.8 is superimposed in Figure 4.14. The overall call success rate is computed as 31.44%, which is slightly lower than that of the security on demand approach. Thus, the increased complexity of NSI does not appear to worsen the call setup time performance appreciably. The key goal of NSI is to disperse the route selection among diverse links of the network. Instead of repeatedly selecting from a limited set of links for different user calls, NSI aims at striking a balance in the even use of all of the network's available resources. Potential advantages include higher reliability, lower vulnerability to natural and artificial catastrophes, uniform degradation of all network resources, and masking network activity. For efficient functioning of NSI, a reasonably rich network interconnection is required, i.e., the network provides multiple alternative paths with adequate available bandwidth between pairs of nodes.

An in-depth analysis of the performance data reveals the following. For 31 of the 40 nodes in the network the call success rates are lower or identical for NSI, relative to security on demand. This is expected, since the NSI approach is likely to choose routes that include more hops and, in turn, compete for available bandwidth at more links. For the remainder of the 9 nodes in the network the call success rate is higher under NSI. While the underlying cause is complex, due to the asynchronous interaction between the nodes and the calls, an examination of the progress of call setup at node 26 reveals the following. At the simulation time instant given by 694,617 timesteps, the bandwidth availability on the link between nodes 26 and 28 is 100 times greater for NSI than for security on demand. This, in turn, permits 5 of the 9 subsequently generated calls at node 26 to be successfully established to and through node 28.

Figure 4.15 presents the average call setup times as a function of the nodes of the ATM network for the NSI augmented security on demand. The plot for the security on demand approach from Figure 4.8 is superimposed in Figure 4.15. The call setup time, averaged over all nodes and calls, is observed at 11.78 ms, which is virtually identical to that in the security on demand approach. In Figure 4.15 the greatest difference in the average call setup times between the two approaches occurs at node 13, and the underlying cause is presented as follows. Under NSI, node 13 claims 24 successful calls, four fewer than under security on demand. However, one of the 24 successful calls is an inter-group call to node 46, which requires a higher call setup time of 33,380 timesteps and is one that security on demand fails to establish. Simulation results reveal that due to the high level of traffic, the available bandwidth at the links is low. This in turn causes a large number of calls to fail. Since the NSI approach tends to select routes with more hops, the increased competition for bandwidth among more links causes a slightly higher failure rate than that of security on demand.

For the baseline, security on demand, and NSI augmented security on demand versions, the average hop counts across all successful calls in the entire network are measured as 4.583, 4.40, and 4.55, respectively. Given that the connectivity

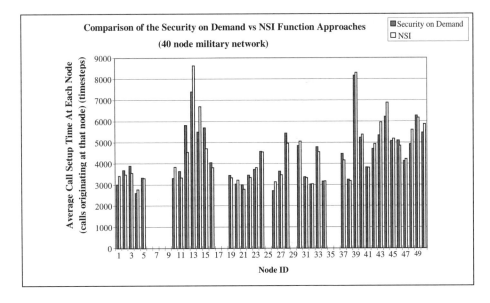

Figure 4.15 Average call setup times for calls originating at each node of the ATM network, for both security on demand and NSI augmented security on demand.

between the groups and within a few of the groups in the representative military network in Figure 4.1 is less than rich, there are fewer alternative routes available for the NSI augmented security on demand version to exploit. Thus, the average hop count value of 4.55 across all successful calls is not appreciably higher than the corresponding values for the baseline and security on demand versions. A closer analysis of the NSI augmented security on demand version, however, yields insight into the influence of NSI on selecting dispersed routes. The remainder of this section presents two examples that constitute careful analysis of the progress of call setup in the actual simulation.

First, a total of 19 call setup requests between Whidbey Island Naval Air Station (NAS) and Fort Lewis in group 3 are examined from the security on demand and the NSI approaches. Figure 4.16 highlights the nodes and topology of group 3. There are two possible routes from Whidbey Island to Ft. Lewis, one via Bangor Submarine Base and another through Seattle, as evident in Figure 4.16. The cumulative physical link propagation delays of the two routes are exactly identical and the bandwidth capacity of each of the links along the two routes is 155 Mb/s.

In the security on demand simulation, the first 8 successful calls are observed to select the route through Seattle, due to the fact that the link between Whidbey Island NAS and Seattle occurs first in the local topology database. For both competing routes, the security posture and the number of hops are identical. During the processing of the ninth successful call, the available bandwidth on the link between Whidbey Island NAS and Seattle drops to zero, and therefore the alternative path through Bangor Submarine Base is selected. By the time the tenth call comes up for processing, the bandwidth availability along the link from

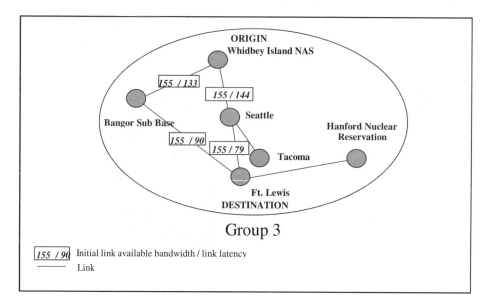

Figure 4.16 Route selection for calls between Whidbey Island Naval Air Station and Ft. Lewis in group 3.

Whidbey Island NAS to Seattle has increased from its previous zero value. Thus, for the tenth call, the route through the Seattle node is selected. While 6 of the 9 subsequent calls are transported via the Seattle node, a total of 3 subsequent calls are routed through Bangor Submarine Base when the bandwidth availability on one or both of the links Whidbey Island NAS to Seattle and Seattle to Ft Lewis again drops to zero.

In the augmented NSI security on demand simulation, the route through Seattle is selected first, since the link between Whidbey Island NAS and Seattle occurs first in the local topology database, and all of the arguments in the NSI are identical for both routes. Thus, for the first 5 successful calls, the route through Seattle is selected. During the processing of the sixth successful call the available bandwidth on the link Whidbey Island NSA to Seattle falls below the threshold of 77.5 Mb/s. Thus, the NSI value associated with this link exceeds that of the link Whidbey Island to Bangor Submarine Base, thereby causing the route through the Bangor Submarine Base node to be selected as shown in Figure 4.17. The next series of calls are routed alternately through Seattle and Bangor Submarine Base, since the NSI values of the corresponding links alternately exceed each other. The final outcome is an even distribution of calls between the two possible paths, implying a uniform dispersion of calls among the possible network resources. In contrast, the security on demand approach appears to favor a particular route until the available bandwidth is completely exhausted.

The second example analyzes call processing at node 41 in group 7 and focuses on the potential limitations of NSI under high stress. Given an inter-group call request to Ft. McPherson in group 5, early in the simulation, NSI causes the se-

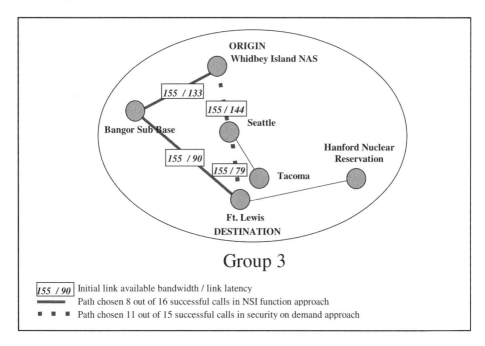

ORIGIN
Whidbey Island NAS

155 / 133

155 / 144

Seattle

Bangor Sub Base

Hanford Nuclear
Reservation

155 / 90

155 / 79   Tacoma

Ft. Lewis
DESTINATION

Group 3

155 / 90	Initial link available bandwidth / link latency
▬▬▬	Path chosen 8 out of 16 successful calls in NSI function approach
▪ ▪ ▪	Path chosen 11 out of 15 successful calls in security on demand approach

Figure 4.17 Route determination for calls between Whidbey Island NAS and Ft. Lewis, under security on demand and NSI approaches.

lection of a route through downtown D.C., and Norfolk, as shown in Figure 4.18. In contrast, security on demand selects a more direct path through Norfolk. Although the route under NSI incurs an additional hop, it is successfully established, since the competition for bandwidth is not yet severe this early in the progress of the simulation.

Next, 4 call requests are launched at node 41 (Alexandria), destined for node 49 (Naval Academy). For the first call request, NSI computes the route through downtown D.C., the Pentagon, and Andrews AFB, as shown in Figure 4.19. However, the call fails due to inadequate bandwidth availability at one of the two highly stressed links Alexandria to downtown D.C., and downtown D.C. to the Pentagon. NSI does not select the more direct route through the Pentagon and Andrews AFB due to its desire to seek a more dispersed route. The NSI approach computes the same route for the next call request, which also fails. Logically, following the failure of the first call request, one would expect the NSI approach to compute a different route for the second call request. However, node 41 updates its NSI table, which contains the NSI values of other links, only when it receives new values through flooding. Also, by design, it recomputes its own NSI values when a call is successfully established and the bandwidth availability is updated. Presumably, between the insertion of the first and second call requests, new NSI values for other links have not been received through flooding, leaving the NSI table for node 41 not current. The NSI approach proposes the same route corresponding to the third call request but this one succeeds, since adequate bandwidth may have

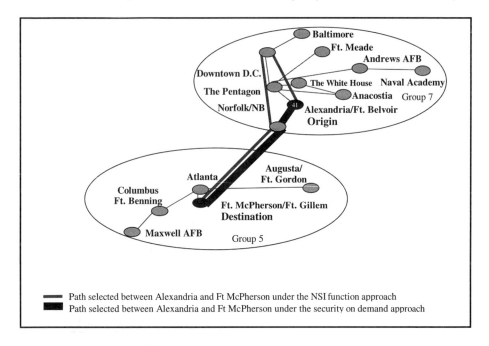

Figure 4.18 Route determination for calls between Alexandria and Ft. McPherson, under security on demand and NSI approaches.

been restored in the two links whose bandwidths were previously depleted. For the fourth call request the NSI approach computes a route through the Pentagon and Andrews AFB to the Naval Academy that proves to be successful. Thus, only

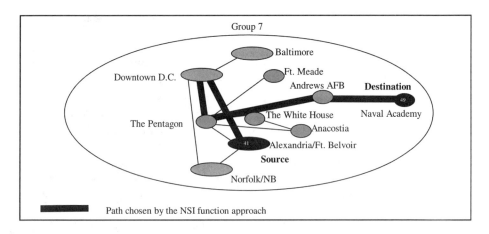

Figure 4.19 Route determination for calls between Alexandria and the Naval Academy, under NSI.

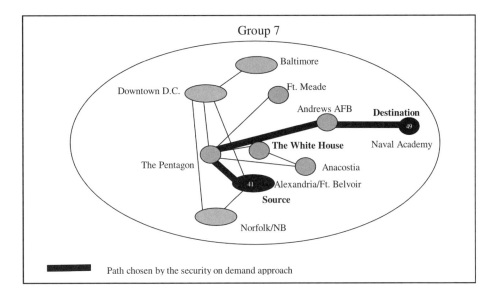

Figure 4.20 Route determination for calls between Alexandria and the Naval Academy, under security on demand.

two of the four call requests succeed, implying a potential limitation of the NSI approach. In contrast, all four calls are successfully established under security on demand, as shown in Figure 4.20, utilizing the direct route through the Pentagon and Andrews AFB, which avoids, inadvertently though, the severely stressed links Alexandria to downtown D.C., and downtown D.C. to the Pentagon.

When incorporated into ATM networks, the security on demand approach enables another exciting application, one where several networks, designed for different security levels under the conventional scheme, may be fused into a single integrated network without compromising the characteristics of any of the individual networks. This is the subject of discussion of the next chapter.

4.6 Problems and Exercises

1. Identify network parameters beyond those discussed in this chapter that may be incorporated into NSI. Justify their role in enhancing the precision of NSI under different input traffic and environmental scenarios.

2. Project with justifications the future trend of the high-level input traffic parameters. What are the implications with respect to the security on demand approach?

5
"Mixed-Use" Network

5.1 Introduction

The current network security paradigm coupled with the desire to transport classified traffic securely has caused the US Department of Defense to maintain its own isolated networks, distinct from the public ATM network infrastructure. Internally, the DoD maintains four types of completely separate and isolated networks to carry Top Secret, Secret, Confidential, and unclassified traffic. A public ATM network may be viewed as carrying unclassified or nonsecure, traffic. While the cost of maintaining four separate network types is becoming increasingly prohibitive to the DoD, the inability of the public and DoD to utilize each other's network resources runs counter to the current atmosphere of dual use and economies of scale. This chapter introduces the concept of a mixed-use network, wherein the four DoD network types and the public ATM network are coalesced into a single unified network that transports all four types of traffic, efficiently and without compromising security. In a mixed-use network the ATM nodes and links that are common to the DoD and public networks are labeled joint-use, and they must necessarily be placed under the jurisdiction of the military for obvious protection of the security assets. This constitutes the first of two key strategies toward the practical acceptance of the notion of mixed-use networks. The control of all other nodes and links remains unchanged. Under the second strategy, although all joint-use links and nodes are subject to military control, the NSI value for a peer node Y recorded at a node X is the result of a new NSI value received from Y through flooding plus other information on the state of Y that X acquires independently through different mechanisms. The concept of mixed-use is the direct result of the user-level security on demand principle that has recently been introduced in the literature and one that is enabled by the fundamental security framework and the basic characteristic of ATM networks.

5.2 Mixed-use Networks: Integrating the Military's Secure ATM Networks with the Public ATM Infrastructure

Due to the obvious risks and the unknown security posture that may result from combining networks, the US military has been unwilling to transport its secure traffic on commercial networks and has continued to build and operate costly separate and totally isolated networks for their secure traffic, distinct from the civilian networks. In addition, even within the DoD, the secure network system is organized into three totally separate networks to carry traffic at (1) Top Secret, (2) Secret, and (3) Confidential levels [22] of classification. The risks include the possible misrouting and/or possible interception of classified data by unauthorized parties, unauthorized use of encryption devices, malicious disruption of unprotected network resources through physical destruction, or denial through remote software bombs or viruses. The current state of the DoD regulations stems from the lack of a proven practical approach to network integration. The security on demand approach enables the military to integrate its three types of classified networks with its unclassified network as well as to include the public ATM network infrastructure in a mixed-use network. Such integration requires that the military assume control of the joint-use nodes and links that are common to the constituent networks, so as to protect their security assets. The control of all other nodes and links remains unchanged. Thus, unclassified DoD traffic may now utilize the additional links and nodes of the civilian network, and the security resources may thereby become more available to serve secure calls. Presently, the DoD does not permit commercial traffic to utilize the links and nodes that it owns exclusively. It is hypothesized that should the DoD implement the security on demand approach, and where the latter yields significant improvement and economy in the use of security resources, the outcome may include a DoD policy change. Commercial traffic requesting secure transport may also be permitted, in a limited manner, to use and pay for the DoD resources or some type of US government-approved security resources.

In this chapter a representative public ATM network is integrated with a representative DoD network to yield a mixed-use network. The representative commercial ATM network shown in Figure 5.1 consists of 32 nodes and 40 links. There is no security present at any of the nodes or links. The network is synthesized utilizing published information on AT&T's [100] and MCI's [101] ConUS commercial backbone networks. The commercial networks generally reveal a rich interconnection topology, implying more alternative paths between the nodes.

The representative military network consists of 40 nodes and 48 links and is shown in Figure 5.2. Identical security resources are present at every link and node and are rated at the highest security level. This also constitutes a continental United States (ConUS) network, and it is synthesized from actual locations of major defense installations and processing centers past and present. Figure 5.2 does not correspond to any known secure military network in existence today. Given a significant number of installations that require connectivity, the presence of multiple alternative types of backup networks, and the desire to reduce costs,

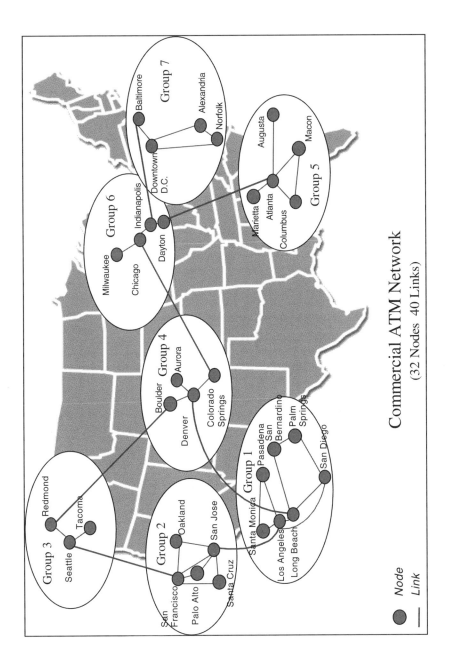

Commercial ATM Network
(32 Nodes 40 Links)

● Node
— Link

Figure 5.1 A 32-node ConUS commercial ATM network.

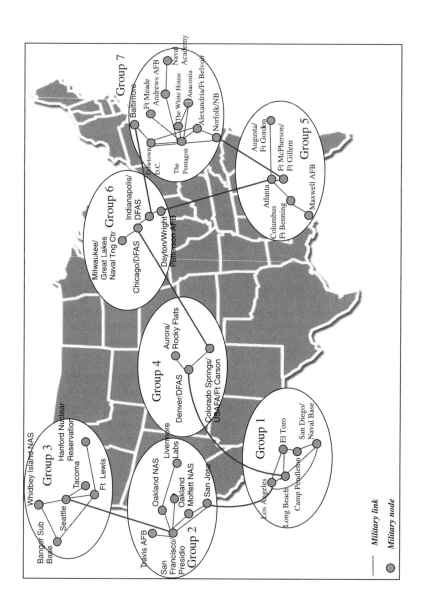

Figure 5.2 A 40-node ConUS military ATM network.

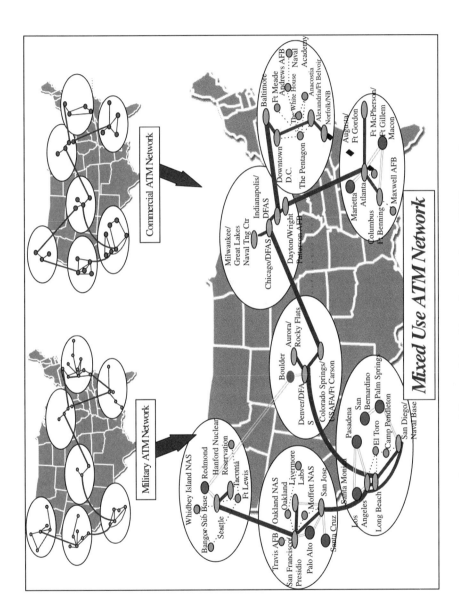

Figure 5.3 The resulting 50-node mixed-use ATM network.

the general rule used in creating the representative network is to limit the number of alternate paths between the nodes.

Figure 5.3 presents a representative mixed-use network that results from the integration of the representative networks in Figures 5.1 and 5.2. It consists of 50 nodes organized into 7 groups and represents a unified network that may be used by all types of users: military, government, industry, and academia. In Figure 5.3 it is noted that while the nodes and links corresponding to the military network in Figure 5.2 preserve their security resources, other nodes and links may lack security. Also, a total of 22 nodes and 23 links overlap both the civilian and military networks and are identified as joint-use in that they will transport both public and military traffic.

In the mixed-use network the military will continue to route its classified traffic through the links and nodes that it owns. However, any unclassified traffic that would previously have been routed over its secure links and nodes is now a candidate for routing over the commercial links and nodes. Thus, the demand for the security resources will diminish, implying higher call success rates for classified traffic.

5.3 Modeling and Distributed Simulation of the Representative ATM Networks

As in the previous chapter, to gain a superior understanding of the behavior of the mixed-use network, each of the three representative networks is modeled on an accurate ATM network simulator developed at ASU and executed on a network of workstations configured as a loosely coupled parallel processor. The simulator consists of over 15,000 lines of code in C/C++, runs on a network of 30+ Intel Pentium workstations under the Linux operating system, and is stimulated by realistic input traffic. Given that it constitutes the key ATM network attribute that enables the notion of mixed-use networks, the model focuses on the call setup process.

5.4 Call Setup in a Mixed-Use ATM Network

As an example of a secure call setup, consider the 9-node network shown in Figure 5.4. In Figure 5.4, while the White House ATM node is connected solely through the Pentagon node, the latter is directly connected to the downtown D.C. node and the Alexandria node, both of which are directly linked to Norfolk. Other military nodes are located at Baltimore, Ft. Meade, Andrews Air Force Base, and the Naval Academy. For every link, its security is encapsulated in detail through a 72-element matrix, as shown in Figure 5.4. Each element of such a matrix may assume a value between 0 and 9, with 0 expressing the most stringent security availability and 9 implying the total absence of security. In addition, for each link, the largest value of any of its 72 matrix elements serves as the top-level security index, which, in turn, is utilized by the augmented call setup process. The top-level security index implies a conservative view and reflects the weakest security item. Consider a

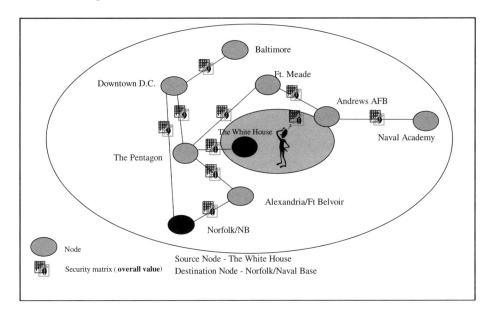

Figure 5.4 Secure ATM call setup between the White House and Norfolk.

scenario where the President, located at the White House node, intends to send a Top Secret message to the Commander in Chief (CinC) of Atlantic Command, located at Norfolk. Clearly, the route selected for this Top Secret message, the only one possible in this network, will include the intermediate nodes Pentagon and Alexandria, since the corresponding links offer top-level security index values of 0. The route is highlighted in Figure 5.5.

As a second example of call setup in a mixed-use network, consider the 9-node network in the greater Baltimore–Washington metropolitan area, as shown in Figure 5.6. In contrast to conventional military networks, where every node and link has identical security resources—an expensive requirement here—while some nodes and links contain the highest security resources, other nodes and links lack any security resource. Thus, the call setup process constitutes a distributed resources allocation strategy, the aim being to allocate security resources to user calls, based upon demand and dictated by need. The nodes in Figure 5.6 include the Naval Academy, Andrews Air Force Base, Ft. Meade, the Pentagon, Baltimore, downtown D.C., Alexandria, Norfolk, and the White House. The security associated with the links between the nodes ranges between Top Secret (0) and Unclassified (9). Consider two military users A and B, both located at the White House and interested in transporting traffic to the Norfolk Naval Base.

While user A insists on a secure route from the White House to the Norfolk Naval Base, user B needs no security. As in the case in Figure 5.5, the secure traffic for user A propagates through the Pentagon and Alexandria intermediate nodes,

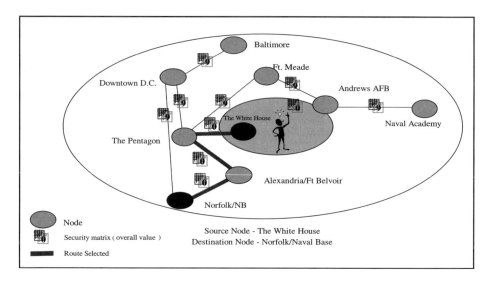

Figure 5.5 Secure route establishment between the White House and Norfolk.

and is shown in Figure 5.7, in black. The unclassified DoD traffic is transported
through the downtown D.C. node, and is shown in Figure 5.7 in gray.

To underscore the benefits of the mixed-use networks, assume that the current
military view is in effect. That is, although A's message requires security, while
B's message requires no security, both messages are treated as classified and are
mandated to be transported along secure links. Thus, in Figure 5.8 both A and B

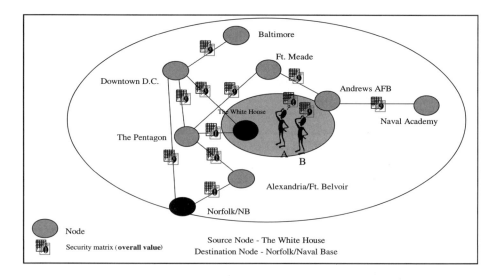

Figure 5.6 A mixed-use network.

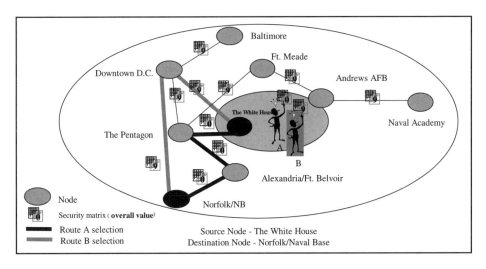

Figure 5.7 Call setup between the White House and Norfolk Naval Base, in a mixed-use network.

compete for the most direct secure route, namely, the intermediate Pentagon and Alexandria nodes. In the event that user B's call setup request precedes that of user A and fully utilizes the available bandwidth, user A's call may fail. Clearly,

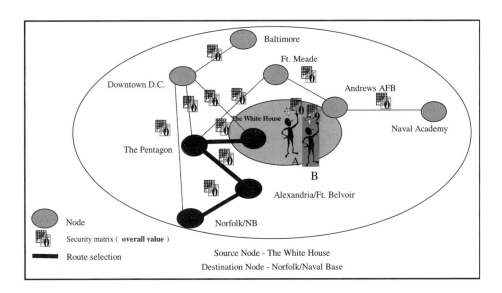

Figure 5.8 Call setup based on the military's current requirement for network security.

this competition is the result of the military's current requirement for network security and is, in essence, unnecessary.

5.5 Modeling the Representative ATM Networks

The 32-node, 40-node, and 50-node representative ATM networks presented in Figures 5.1, 5.2, and 5.3, respectively, are modeled using an accurate ATM simulator. Figure 5.9 represents the 50-node mixed-use network along with the node and group identifiers. Key issues in the modeling include the following. The behavior of each ATM call processing node, including the ATM Forum's PNNI specification, is encapsulated through a UNIX process. The links between the ATM nodes are represented through TCP-IP connections between the corresponding processes, and the transport of the call setup messages is modeled through guaranteed messages between the processes. The processes are executed asynchronously, utilizing a conservative, null message based, asynchronous distributed simulation algorithm [93]. The computing testbed consists of a network of 30+ Pentium workstations under the Linux operating system, configured as a loosely coupled parallel processor. Thus, the ATM simulator is distributed and unique in that it closely resembles an operational ATM network. Furthermore, fast and accurate simulation of large-scale networks is permitted, enabling a systematic study of the performance impact of security on demand for different choices of network parameters and input traffic.

In the 32- and 40-node ATM networks, the links between the nodes that are labeled joint-use in Figure 5.9 are rated at one-half of 155 Mb/s, or 77.5 Mb/s. All other links are rated at 155 Mb/s. When the 32- and 40-node ATM networks are coalesced into the 50-node mixed-use network, the overlapping joint-use links combine to yield a link of net rating 155 Mb/s. The sole commercial and military links remain unchanged, and thus all links are rated at 155 Mb/s for the 50-node mixed-use network. For every link, the propagation delay is computed by obtaining the straight-line distance between the nodes and dividing it by the speed of light in fiber, 194,865 km/sec [92]. While the security index values of the links in the military network are set to 0, implying the availability of the best security resource, those associated with the public network are set to 9, implying no security. In the mixed-use network, the security index values for the joint-use links are set to 0, as are the links that occur in the military network. For all other links, the security index value is set to 9.

5.6 Simulation Experiments and Performance Analysis

The principal objective of this section is twofold: (1) to demonstrate scientifically the feasibility of the concept of mixed-use networks, and (2) to analyze the benefits of integrating the military and public ATM networks, if any, into a unified mixed-use network. The first objective is achieved by executing the 50-node, mixed-use network, modeled for security on demand, on the testbed and verifying that secure and nonsecure routes are computed for classified and unclassified traffic,

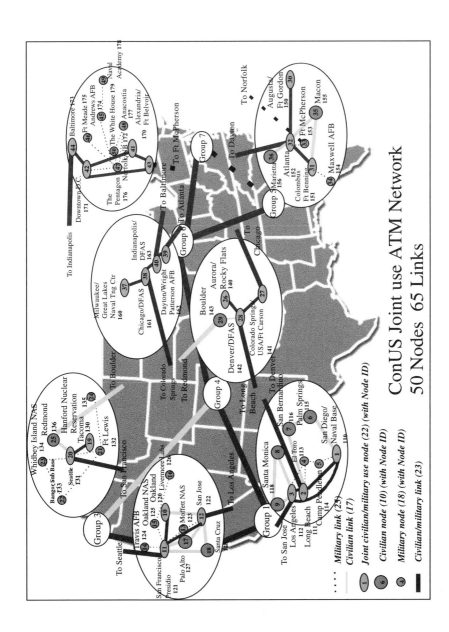

Figure 5.9 The 50-node mixed-use ATM network, with node and group identifiers.

respectively. For the second objective, this research focuses on two metrics: success rates of call requests and call setup time. While the call success rate measures the number of calls for which routes are successfully obtained through the network, the call setup time measures the total time, in timesteps, for the network to establish a route for a successful call.

5.6.1   Metric I: Call Success Rate

Figure 5.10 presents the call success rate for every node in each of the three representative networks, averaged over the entire simulation time period. While data points occur for every node for the 50-node mixed-use network, data points may be lacking for specific node identifiers in the 32- and 40-node networks that do not exist in the corresponding networks. Visual inspection of Figure 5.10 reveals that the call success rates are higher across the nodes of different groups for the mixed-use network than for the military network, except for group 7. The slight decrease in the call success rate for the nodes in group 7 in the mixed-use network may be due to the increased volume of public calls, inserted at the joint nodes 41, 42, 43, and 44. The public calls compete with the DoD calls for bandwidth through the links between nodes 42 and 47 and nodes 47 and 41, and cause the latter calls to fail. Node 47 constitutes a critical node in the routes to all of the solely military nodes within group 7.

The inference of higher call success rates for the mixed-use network is supported by the overall call success rate, i.e., over the entire network, of 40% for the mixed-use network as opposed to 31.4% for the military network. The actual numbers of successful calls are 1145, 950, and 2101 for the public, military, and mixed-use networks, respectively. Clearly, the total number of successful calls in the mixed-use network, 2101, is slightly higher than 1145 + 950 = 2095, the combined total number of successful calls for the public and military networks, implying that the richer interconnection topology of the mixed-use network may have given rise to multiple alternative routes between the nodes. Analysis also reveals that the total number of successful secure DoD calls originating within group 2 in the 40-node network is 24. For the 50-node mixed-use network, the total number of successful secure DoD calls originating within the same group 2 sharply jumps to 45. This behavior is representative throughout all of the groups in the mixed-use network.

As expected, a weakness of the mixed-use network is the potential for a drop in the call success rate for the public ATM network, caused by the DoD's use of the public network to transport unclassified traffic. Figure 5.10 confirms this limitation in that the overall call success rate drops from 51.9% for the public network to 40% for the mixed-use network. Conceivably, the advantages that may offset this weakness include the possible infusion of infrastructure by the military into the public network and the ability for select public users to utilize DoD's security resources, if and when permitted, to transport critical information.

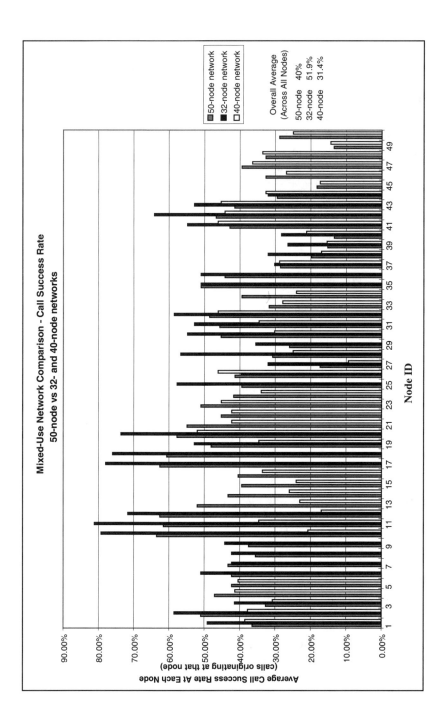

Figure 5.10 Comparative call success rates for the representative public, military, and mixed-use networks.

5.6.2   Metric II: Call Setup Time

The time required to establish a successful call in an ATM network is a function of the physical link propagation delays and node delays, which, in turn, are dependent on the processing delays at the nodes and any buffer delays arising from the need to buffer the call requests. In the behavior model for an ATM node, a call processing delay of 2.192 ms is introduced, which corresponds to the computation time required for every call. The reasons underlying the choice of 2.192 ms have been addressed earlier, in section 4.5.

Figure 5.11 reports the call setup times incurred at every node of the three representative networks, averaged over all calls originating at the node. Analysis reveals that generally, all nodes of the mixed-use network incur a larger call setup delay than the public and military networks. The average call setup delay of 6736.1 timesteps over all calls in the mixed-use network exceeds that in the public network by 2867.8 timesteps and in the military network by 2436.9 timesteps. The reasons are as follows. First, in each of the 32- and 40-node networks, approximately one-half of the calls are originated at the joint-use nodes. Therefore, relative to the joint-use nodes in each of the 32- and 40-node networks, those in the 50-node network are subject to twice the number of calls, thereby implying higher values for the call setup time. The second reason is more subtle and is explained as follows. In each of the 32- and 40-node networks, a certain subset of the calls in each of the systems, generated at other than the joint-use nodes, will require processing by the joint-use nodes. In the 50-node network, the joint-use nodes are likely to process both of these subsets, implying an increase in the average call setup time. Given the choice of the 50-node network topology wherein the joint-use nodes intercept most of the routes through the network, the result is longer average call setup time across all nodes. Third, select nodes in the mixed-use network may incur additional links from other nodes, in contrast to the topology in the original public or military networks, implying an increase in the number of calls that it must process. For example, nodes 26, 27, 28, and 29 of group 4 in the mixed-use network incur an extra call processing burden in contrast to the military network, because of the additional link from group 3. Thus, while the average call setup delay may increase, the chances for previously unsuccessful inter-group calls to succeed are also higher. The impact of this characteristic of the network topology on the value of the average call setup time for nodes 26 through 29 is nonlinear and significant, as revealed in Figure 5.11, and it skews the value of the average call setup delay over all calls. Ignoring the influence of the nodes of group 4, the average call setup delay over all calls computes to 5337.6 timesteps, while ignoring the influence of both groups 4 and 6 yields a value of 5088.8 timesteps.

A careful examination of an actual call between the Seattle node in group 3 and the Rocky Flats node in group 4 is helpful in understanding the general behavior that occurs throughout the network. Figure 5.12 tracks the progress of a call originating from Seattle in group 3 in the military network. The call finds a route through San Francisco and San Jose in group 2 but fails in group 1 at Los Angeles due to lack of bandwidth. If the call had been successful, the route would have

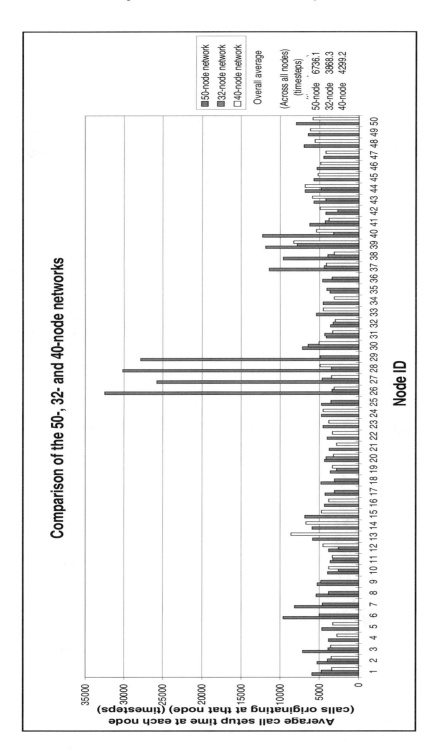

Figure 5.11 Comparative call setup times for the representative public, military, and mixed-use networks.

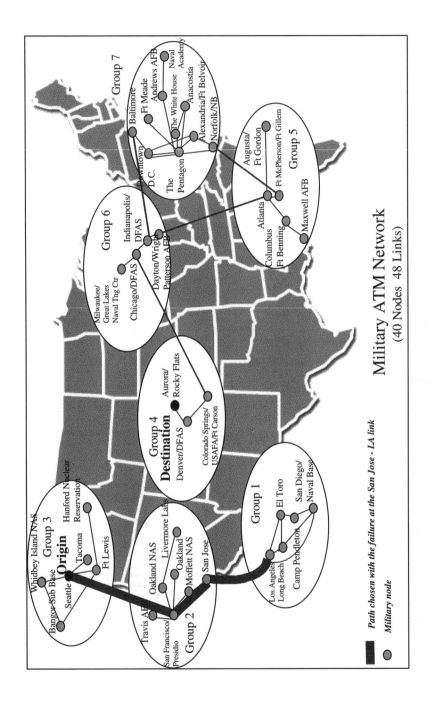

Military ATM Network
(40 Nodes 48 Links)

Path chosen with the failure at the *San Jose - LA link*

● *Military node*

Figure 5.12 Progress of a call from Seattle to Rocky Flats in the military network.

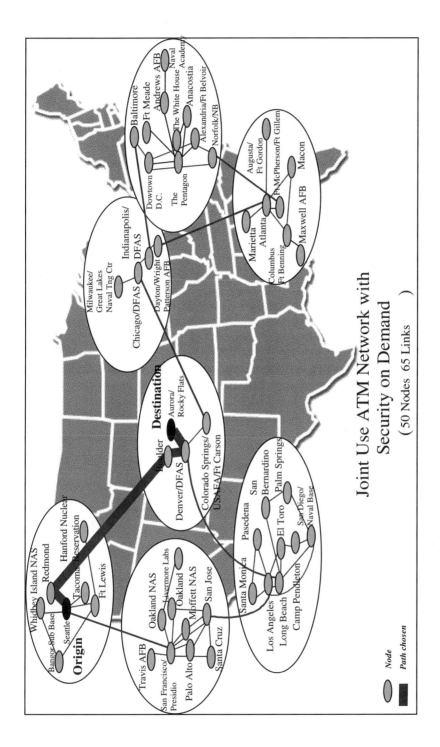

Figure 5.13 Progress of a call from Seattle to Rocky Flats in the mixed-use network.

included the intermediate nodes LA, Long Beach, and Denver, implying a total of 6 hops to the final destination.

The mixed-use network inherits a link between groups 3 and 4 that is otherwise missing in the military network. A trace of the same call in the mixed-use network, as shown in Figure 5.13, reveals that the call succeeds, and the route includes the intermediate nodes Redmond, Boulder, and Denver, for a total of 4 hops. For the unclassified DoD call requests, the mixed-use network is expected to offer a richer connectivity, in contrast to the military network, implying a potential reduction in the average number of hops encountered by the calls in the mixed-use network. Measurements confirm this expectation. For the military network, the total number of successful calls is 4,329, while the cumulative hop count for all successful calls is 10,443, implying an average hop count of 2.41 per call. For the mixed-use network, the number of successful calls is 7,948 and the cumulative hop count 15,585, yielding an average hop count of 1.9 per call.

5.7   Problems and Exercises

1. What implicit assumptions underlie the success of mixed-use networks?

2. What are the implications of generalizing the mixed-use concept to an underlying nationwide infrastructure?

6

Systematic Analysis of Vulnerabilities and Synthesis of Security Attack Models for ATM Networks

As complex systems, networks consist of a number of constituent elements that are geographically dispersed, are semiautonomous in nature, and interact with one another and with users, asynchronously. Given that the network design task is already intrinsically complex, it is natural for the traditional network designer to focus, in order to save time and effort, only on those principles and interactions, say D, that help accomplish the key design objectives of the network. The remainder of the interactions, say U, are viewed as "don't cares" or passive, bearing no adverse impact under normal operating conditions. In reality, however, both internal and external stress may introduce abnormal operating conditions into the network under which the set U may begin to induce any number of unintended effects, even catastrophic failure. A secure network design must not only protect its internal components from obvious attacks from the external world, but, and this is equally important, resist internal attacks from two sources, foreign elements that successfully penetrate into the network and attack from within and one or more of the internal components that spin out of control and become potentially destructive. This chapter introduces the notion of network vulnerability analysis, conceptually organized into three phases. Phase I focuses on systematically examining every possible interaction from the perspective of its impact on the key design objectives of the network, and constitutes an indispensable element of secure network design. Given that the number of interactions in a typical real-world network is large, to render the effort tractable, phase I must be driven from a comprehensive and total understanding of the fundamental principles that define the network. Phase I is likely to yield a nonempty set of potential scenarios under which the network may become vulnerable. In phase II, each of these weaknesses is selected, one at a time, and where possible, a corresponding attack model is synthesized. The purpose of the attack model is to manifest the vulnerability through an induced excitement and guide its effect at an observable output. The attack model assumes the form of a distinct executable code description, encapsulating the abnormal behavior of the network, and assumes an underlying executable code description that emulates the normal network behavior. In phase III, the attack models are simulated, one at a time, on an appropriate testbed, with two objectives. First, the simulation verifies the thinking underlying the attack model, i.e., whether the attack model succeeds in triggering the vulnerability and forcing its manifestation to be detected at an observable output. When the first objective is met, the simulation often reveals the impact of the attack model on network performance. Under the second objective, the extent of the impact is captured through an innovative metric design. The idea of vulnerability analysis closely resembles the techniques of fault simulation and test generation in the discipline of computer-aided design of integrated circuits (ICs). Under fault simulation, to detect the presence of faults in a manufactured IC, first a fault model is proposed, reflecting the type of suspected failures, and second, the IC is "fault simulated" to flush out as many of the internal faults as possible at the observable outputs.

This chapter focuses on a vulnerability analysis of ATM networks and is organized into three parts. Part 1 is a brief review of the current literature on attack models. Part 2 presents the fundamental ATM network principles that form the basis for launching the vulnerability analysis. Part 3 focuses on synthetic attack models, their detailed assessment through modeling and asynchronous distributed simulation on ATMSIM 1.0 [94], and the lessons learned for hardening ATM networks in the future.

6.1 Brief Review of the Current Literature on Attack Models

A careful examination of the literature on attack models reveals that nearly all efforts focus on data networks. The literature also reveals the lack of a systematic analysis of the vulnerability of data networks from basic principles. As a result, both a set of basic attacks derived from the vulnerability analysis and an approach to validate such attacks are missing in the literature. Virtually all of the reported efforts, except for [68], are ad hoc, lack systematic modeling and analysis, and do not support efforts to use them to harden networks. Since data networks including the Internet have been around the longest, the literature contains a rich set of documented methods to launch attacks on the Internet as well as a collection of clever techniques to defeat them.

6.1.1 Denial of Service Attack via PING

A denial of service attack via PING occurs when oversized IP packets are transmitted inadvertently or intentionally via PING. Given that the maximum packet size in TCP/IP is set at 65,536 bytes, when confronted with oversized IP packets, networks may react in unpredictable ways including crashing, freezing, or rebooting of the system. Two straightforward remedies are:

- Appropriate handling of large packets.

- Reject oversized ICMP datagrams.

6.1.2 Password Breaking

Though crucial to system security, system files such as /etc/passwd are normally readable by all users, tempting perpetrators to attempt to break the encrypted information stored in these files. A possible solution consists in utilizing shadow passwords.

A key vulnerability for any system is the possible compromise of the root password, especially when it remains unchanged over a long period of time. A possible remedy consists in using one-time passwords for the root account. Techniques to generate a one-time password include:

- Synchronization between the computer and an electronic card carried by the user.

- A challenge and response method.

6.1.3 TCP Wrapper

The proposed enhancement permits the system administrator to selectively reject requests for services such as ftp, rsh, and rlogin from suspicious sources. Specific measures may include:

- Denial of specific services from outside a list of specified domains.

- Rejection of all requests for services from a particular host identified as hostile.

6.1.4 Data Encryption

As a logical step, sensitive data may be encrypted to defeat unauthorized tapping and access by intruders. The techniques may be further classified as follows:

- Encrypting Data On Disk: The user assumes responsibility for encrypting any sensitive information stored on the disk. Despite the obvious benefits, the drawbacks include the following:

  - Requires additional processing time to encrypt and decrypt the data.

  - The encrypted data might become inaccessible and thus lost if the encryption key is misplaced.

- Encrypting Data in the Network: Prior to launching it on the network, sensitive data is encrypted to prevent unauthorized computers with access to it from reading and manipulating it. For example, a computer on an Ethernet link may exploit the promiscuous mode of Ethernet cards to read data that is intended for a different computer. Limitations of this approach include:

  - Encryption slows down the process of propagating data over the network.

- Distribution of the encryption key to authorized recipients without allowing a hacker grab it first is a difficult problem. A possible solution consists in employing a set of public and private keys.

6.1.5 Firewalls

The philosophy underlying firewalls is to shield the network from the external world and concentrate attacks at a single point that can be effectively monitored by the network management. The advantages and limitations are as follows:

6.1.5.1 *Advantages.*

- All possible attacks on the system are guided to a single point: the firewall.

- Additional protection and strategic planning may be accomplished through monitoring the firewall activity.

6.1.5.2 *Limitations.*

- Firewalls may constitute a false sense of security in that they do not protect the network against insider threats, documented as the the biggest source of threat.

- In a firewall situation, all security resources are focused at the firewall, leaving the internal components unprotected and vulnerable.

- Firewalls tend to isolate users, denying them the ability to utilize the network resources completely and interfering with the quality of their work.

- Often, the restrictions imposed are so severe that users demand that holes be inserted in the firewall to permit key services, at the cost of greatly diminished protection by the firewall.

- The industry estimates the reduction in performance stemming from the use of firewalls at approximately 90%.

6.1.6 Trojan Horse

The concept of Trojan horse attacks refers to the technique of secretly implanting hostile entities in a network. Thus, in theory, unless a system is designed and implemented completely by one trusted individual, the system does not and should not lend itself to 100% trust. Thompson [68] warns of the danger of compiling malicious code, deliberately or accidentally, into the operating system, labeling them Trojan horses. A careful analysis reveals that all networks are fundamentally vulnerable to the metastability problem [94][11] that affects every sequential digital design when a flip-flop encounters an asynchronous input signal from the external world.

6.1.7   Classification of Attacks on Networks in the Literature

The literature organizes potential security violations into three distinct categories:

- Unauthorized release of information

- Unauthorized modification of information

- Unauthorized denial of resources

All of these attacks may be further classified into two types, active and passive, described below.

- Active attacks are capable of modifying information or causing denial of resources and services. Active attacks are further subdivided into two categories:

  - Message-stream modification attacks that affect the authenticity, integrity, and ordering of the protocol data units (PDUs) propagating through the network. Attacks on authenticity may be realized by either modifying the protocol control information in the PDUs such that they are sent to the wrong destination, or through inserting bogus PDUs into the network. Attacks on integrity are accomplished by modifying the data portion of PDUs, while attacks on ordering are realized through deleting PDUs or modifying sequencing information.

  - Under denial of message service attacks, the intruder either discards all PDUs propagating along the network or simply delays all PDUs in one or both directions.

- Under passive attacks, the intruder merely observes the passage of the PDUs on the network without interfering with their flow. Thus, passive attacks simply cause information release in that the intruder may examine the protocol control information portion of a PDU, recording the location and identities of the communicating protocol entities. Thus, the mechanism underlying passive attacks is similar to that of traffic analysis.

6.1.8   Recommendations on Mechanisms to Provide Communications Security

To effectively address the different types of attacks described earlier, the literature recommends that the following objectives be addressed:

- Prevention of release of message contents

- Detection of message-stream modification

- Detection of denial of message service

- Detection of spurious association initiation

6.1.9   Proposed Approaches to Communications Security

To provide communications security, the literature recommends two basic techniques, link-oriented and end-to-end security measures, as described subsequently.

- Under the rubric of "link-oriented," the thinking is to provide security to the information being transmitted over an individual communication link between two nodes, regardless of the ultimate source and destination of the information. Since the information is encrypted only on the link, the nodes at either end of the link must necessarily be secure. The weaknesses are:

  - A single nonsecure intermediate node can expose a substantial amount of message traffic, despite extensive and expensive link encryption.
  - Cost of maintaining security in the links and nodes, on a continuing basis, can be very high.

- Under "end-to-end," the thinking is to provide uniform protection to every message propagated from the source to the destination, regardless of whether the intermediate nodes and links are secure. Thus, the key benefits are:

  - An individual user or host node may choose to propagate a message securely without affecting other users and host nodes.
  - In addition to packet-switched networks, the end-to-end technique may be useful in packet-broadcast networks where link encryption is not always available.

6.2   Fundamental Characteristics of ATM Networks

The ATM networking technology was designed by the International Telecommunication Union Telecommunication Standardization Sector (ITU-T), intended as a standard for the high-speed transfer of voice, video, and data through public and private networks. The ATM design reflects the desire to integrate the benefits of circuit switching, namely, guaranteeing resources and constant transmission delay, with the advantages of packet switching: flexibility and efficiency for highly dynamic, intermittent, asynchronous traffic. Current standardization efforts for ATM are led primarily by the ATM Forum.

6.2.1   Key Characteristics

ATM differs from all other networking strategies through the combination of the following three characteristics:

- Fixed-size cells: All traffic—audio, video, or data—is organized into fixed-length packets to permit greater efficiency in switching, from the hardware perspective, than for variable-length packets.

- Connection-oriented service: All of the cells of a given message are propagated along the same route that is established following a connection request and prior to launching the first traffic cell.

- Asynchronous multiplexing: To ensure fairness and efficient use of bandwidth resources, cells arriving at the input of any ATM switch, possibly from different users, are selected through statistical multiplexing, subject to priority considerations, and processed by the switch.

By combining these characteristics, ATM can provide a number of different categories of service for users with a wide range of data requirements. Following a user's request, a service contract is established and a virtual connection is set up from the source all the way to the destination. As a consequence of the contract, the connection guarantees a specific bandwidth and other QoS parameter values throughout the life of the connection.

6.2.2   ATM Layers

To understand the packetization and transport of traffic in ATM, consider Figure 6.1, which describes the key layers. Though not an inherent part of the ATM network, the ATM adaptation layer, AAL, is an important standardized interface between the ATM network and upper layers. AAL is organized into two sublayers, SAR (segmentation and reassembly) and CS (convergence sublayer). At the

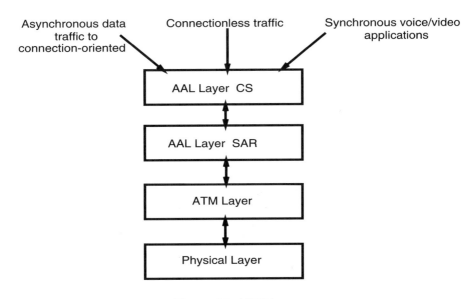

Figure 6.1 ATM layers.

source, SAR processes user data units of different sizes, formats them into ATM cells, and forwards them onto the ATM layer for processing by the ATM switch. At the destination, SAR reassembles the ATM cells and reformats them back into the user-specified format. The function of the CS is dictated by the type of traffic—voice, video, or data—being processed by the AAL. The ATM layer is connection-oriented, so cells must be associated with established virtual connections. The underlying ATM switch employs VPI/VCI labels, explained subsequently in section 6.2.3.1.1 to identify the connection with which each cell is associated.

6.2.3   ATM Network Interfaces

ATM devices are interconnected over point-to-point links through two types of interfaces: user network interface (UNI) and the network–network interface (NNI). As shown in Figure 6.2, while an ATM switch, also referred to as an ATM node here, is connected to another ATM switch through NNI, an ATM switch is connected to an ATM end system, representing the users, through the UNI.

6.2.3.1   *ATM Cell.* An ATM packet is exactly 53 bytes in size with the cell header occupying the first 5 octets and the payload contained in the remaining 48 bytes, as shown in Figure 6.3. The thinking underlying the choice of fixed packets is to permit fast hardware processing, while that behind the choice of relatively small packet size is higher efficiency in switching and desire for lower transmission delays.

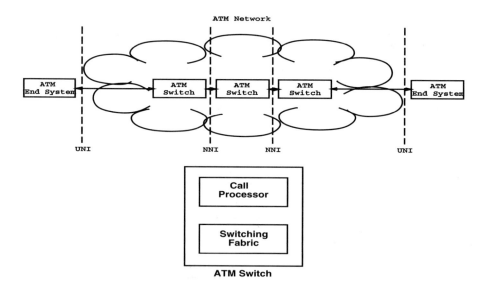

Figure 6.2 ATM network topology.

Figure 6.3 ATM cell.

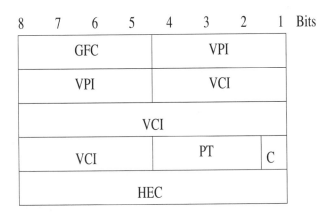

Figure 6.4 ATM cell header: UNI format.

6.2.3.1.1 *Cell Header Format*

The 5-byte, ATM cell header assumes different forms depending on whether the underling cell is a resident of the UNI or NNI interface. Figure 6.4 presents the organization of the UNI cell header.

- Generic flow control (GFC): Provides local functions including flow control between the ATM end system equipment and the ATM switch.

- Virtual path identifier (VPI) and virtual channel identifier (VCI): VPI and VCI together identify a virtual channel on an ATM link. A concatenation of such channels through the switches constitutes a virtual path connection (VPC) or virtual channel connection (VCC) across the network.

- Payload type (PT): The first bit of PT indicates whether the cell contains user data or control information. Where the cell contains user data, the second bit of PT identifies whether congestion is encountered, while the third bit reveals whether this cell is the last one in a series of cells that represent a single AAL5 frame. Where the cell contains control information, the second and third bits indicate maintenance or management flow information.

- Cell loss priority (CLP): Where the CLP value of a cell is set to 1 and the cell encounters heavy congestion, the cell may be dropped. Otherwise, where

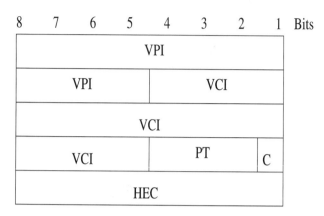

Figure 6.5 ATM cell header: NNI format.

the CLP is set to 0 and the cell encounters heavy congestion, the system may discard the cell.

- Header error control (HEC): Consists of a cyclic redundancy check on the cell header.

Figure 6.5 presents the format of the NNI header, which is identical to that of the UNI header, except for the following. The GFC field is eliminated, and more bits are allocated to the VPI, taking it up to 12 bits and making more VPIs available for the network–network interface.

6.2.3.2 *Virtual Paths and Virtual Channels.* In ATM networks, virtual connections come in two forms:

- Virtual path connections, identified by virtual path identifiers (VPIs).

- Virtual channel connections, identified by the combination of a VPI and a virtual channel identifier (VCI). A virtual channel approximates a classic virtual circuit.

As seen in Figure 6.6, a virtual path is a bundle of virtual channels, all of which are switched transparently across the ATM network on the basis that they share a common VPI. A virtual path connection is a collection of virtual channels.

6.2.3.3 *ATM Switch.* Conceptually, an ATM switch is organized into two elements: a switch fabric and a call processor. Figure 6.7 elaborates on the functions of the ATM switch. The organization for establishing standards, ATM Forum, has proposed two protocols: PNNI, the private network-to-network protocol, and UNI, the user–network interface protocol. While PNNI is for use between private ATM switches and between groups of private ATM switches, UNI lays the communications groundwork between a user and the interfacing ATM switch of the ATM network.

Figure 6.6 Virtual paths and virtual channel connections.

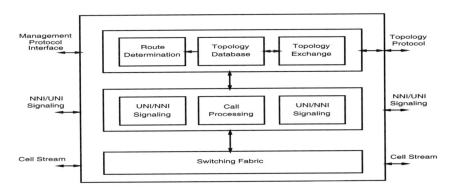

Figure 6.7 Reference model for an ATM switch.

- Under PNNI, continuously updated topology information is distributed between switches and clusters of switches, which is used to dynamically compute paths through the network. A hierarchical organization attempts to ensure that the protocol is scalable for large worldwide ATM networks. PNNI topology update and routing are based on the well-known link-state routing technique.

- Included in UNI is a signaling protocol that is used to establish point-to-point and point-to-multipoint connections across an ATM network. This protocol has added mechanisms to support source routing, crankback, and alternative routing of call setup requests in case of connection setup failure.

6.2.4   Call Establishment and Call Clearing

Figures 6.8 and 6.9 summarize the procedures for call establishment and call clearing. Call establishment is initiated by the terminal equipment sending a SETUP message. Of the different fields in the SETUP message, the most important is the call reference, a unique number that identifies the call. SETUP also contains a route from the source node to the destination node, specified in the DTL-designated transit list field. The entire call establishment process includes the following message types:

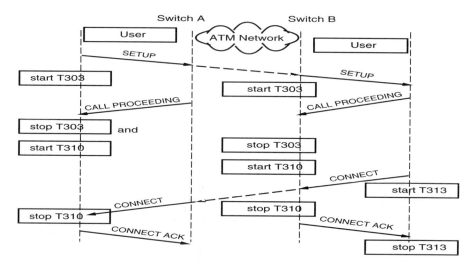

Figure 6.8 Successful call establishment.

- **SETUP**: To initiate and propagate a call establishment request.

- **CALL PROCEEDING**: To acknowledge receipt of SETUP and to indicate that the call request is being processed.

- **CONNECT**: To indicate that the call has been successfully established between the source and destination nodes.

- **CONNECT ACKNOWLEDGMENT**: To acknowledge the CONNECT message.

- **RELEASE**: To request rejecting or disconnecting a call.

- **RELEASE COMPLETE**: To acknowledge a RELEASE message, after the necessary steps have been undertaken to release the call.

Associated with the message types mentioned earlier are timers T303, T310, and T308, which are described below:

- **T303** is initiated when an ATM node sends a SETUP message to the subsequent node that is specified in the DTL field of SETUP. It is reset when the ATM node receives a CALL PROCEEDING message.

- **T310** is started after a CALL PROCEEDING message is received and discontinued upon receipt of a RELEASE, CONNECT, or RELEASE COMPLETE message.

- **T308** is initiated when a RELEASE message is sent out by a node and reset when the node receives a RELEASE COMPLETE message.

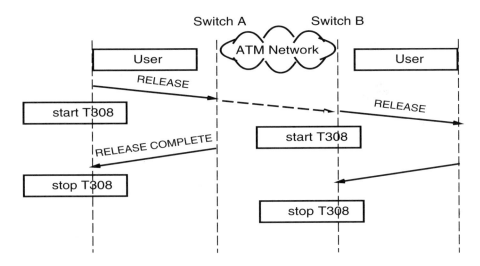

Figure 6.9 Successful call release.

Upon expiration of any of the timers, a corresponding message is sent out once again and the timer is restarted. Upon final expiration of the timer, the call clearing procedure is initiated under the cause "recovery on timer expiry."

6.2.5 Mirror-Image Network

Conceptually, an ATM network is composed of two subnetworks, one an identical mirror-image of another. In Figure 6.10 the subnetwork in the middle refers to the switching fabric network, where the nodes are the switch fabrics and the links carry the user traffic, while that at the bottom represents the call processing subnetwork, where the nodes, the call processor engines, are interconnected through signaling links. The composite network is represented by the diagram at the top of Figure 6.10.

6.2.5.1 *Call Processor and Switch Fabric Operations.* Upon receipt of an incoming cell, an ATM switch first determines whether it is of a traffic or signaling type. Signaling cells are sent to the call processor, which assumes responsibility for connection setup and release. Traffic cells are forwarded to the switch fabric where they are queued in the input buffer. The cell header is examined to extract the VPI/VCI pair, which is then utilized to redirect the cell on the intended outgoing channel. The corresponding output VPI/VCI values are read from the VPI/VCI mapping table, written into the updated cell header, and the cell is forwarded to the output buffer queue for subsequent propagation along the appropriate outgoing port.

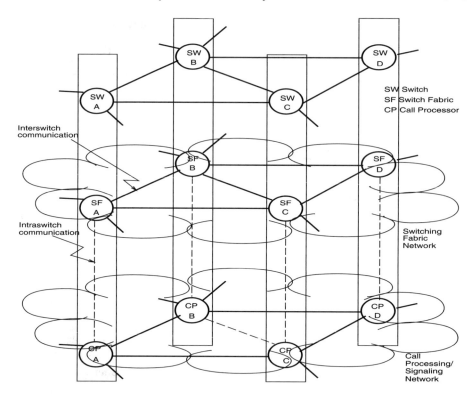

Figure 6.10 ATM network composed of two mirror-image subnetworks.

6.2.6 Summary

This section has presented a brief operational description of the ATM network, the intent being to analyze it and uncover any weakness that may be exploited by smart perpetrators to successfully attack the network. The analysis and synthesis of attack models are presented in the next section. The eventual goal is to build appropriate hardware and software countermeasures or, if necessary, even redesign the ATM network principles to defeat such attacks.

6.3 Synthesis of Security Attack Models for ATM Networks

An attack may be viewed as a perturbation of an operationally correct ATM network. Given the complex nature of ATM networks, conceivably, a rare combination of user traffic and system parameters may automatically lead to an adverse interaction between the different ATM principles, reflecting the lack of comprehensive thinking during design. This section, however, will be confined to attacks that are deliberately planned by intelligent enemy agents, starting from a thorough analysis of the ATM vulnerabilities. The vulnerability analysis effort will build on the fundamental framework for network security presented in Chapter 2, and focus on the four key components of ATM networks: switch fabric, call processor, ATM links, and the basic ATM operating principles including PNNI and UNI. Once a vulnerability has been identified, the goal is to synthesize an attack that will stimulate the vulnerability and cause it to manifest itself in the form of perceptible performance impacts. The attack may consist of a contrived set of user interactions with the network, user traffic, and system parameters that is generated synthetically. In this section the focus is to identify complex attacks that while based on the ATM fundamentals are representative of those that would be constructed by thoughtful enemy agents. Attacks are organized into two broad categories. The first attack type, basic, focuses on the failure of specific standard functions in ATM networks, while the second attack type is labeled complex, and it refers to the prescription of a malicious intent or objective. Highly sophisticated attacks including timing attacks and those coordinated across different geographical points are conceivable. These may be extremely difficult to detect and defeat and are not discussed in this book.

6.3.1 Vulnerabilities

6.3.1.1 *Unrestricted User Interactions with UNI.* A significant strength of ATM networks, as stated earlier in section 6.2, is the lack of restrictions imposed on the users in their different interactions with the network. First, a single user may request any number of different call requests at the UNI, either for a single destination node or different destination nodes. Second, a user may request any combination of the QoS traffic parameters. Although any call request is approved only when appropriate network resources are available, ATM's permissiveness may be successfully exploited by perpetrators.

6.3.1.2 *In-band Signaling and Nonseparation of Resources for Traffic and Signaling Cells.* Given the significant bandwidth of ATM links, 155 Mb/s and higher, and the relatively modest bandwidth requirements of the signaling network, ATM Forum's decision to use the switching fabric network shown in Figure 6.10 to transport signaling cells, instead of constructing a separate signaling network, is logical and makes economic sense. However, under these circumstances, termed in-band signaling, both the signaling and traffic cells have access to a key resource within the ATM switch, namely buffers. This may construe a serious vulnerability, limited by the imagination of the perpetrator. The use of in-band signaling was popular in the early Bell System telephone networks, around the time of World War II, primarily for efficiency and economy. No additional bandwidth was required, and the signaling could easily be moved across available speech channels. However, clever users soon discovered mechanisms to gain access to the switch by sending special characters during a conversation session, and they could dial long-distance calls without being detected [102][103]. This cost Ma Bell untold dollars [104]. Other mischief is equally conceivable. Eventually, in-band signaling was disbanded in favor of out-of-band signaling, with the creation of a signaling network completely isolated from the voice network for reliability and security.

6.3.1.3 *Unrestricted Access to Cell Headers.* By definition, every intermediate switch must possess at least complete read access to every cell header. The header contents must be analyzed to yield the subsequent path for the cell. This property of ATM switches may constitute a vulnerability.

6.3.1.4 *Vulnerability of VPI/VCI-Based Switching.* Following a successfully established connection between the source and destination nodes, corresponding to a user call request, user traffic is transported along the channel in the format of 53-byte ATM cells. Each cell contains a 5-byte header containing a VPI/VCI pair, which, coupled with the VPI/VCI translation table entry created at the intermediate ATM switches at the time of call establishment, ensure the correct switching of the cells along the predetermined channel from source to destination. ATM Forum's PNNI document does not clarify whether the call processors at the intermediate switches must retain the source node and destination node addresses, call reference number, etc. that are specified in the DTL and utilized while the call SETUP request is being processed. Thus, according to current standards, only the VPI/VCI values control the routing of user traffic cells, which renders the VPI/VCI pair sensitive and vulnerable.

6.3.1.5 *Vulnerability of the Call Processor.* The call processor plays a major role in determining whether a user's call request is to be accepted or rejected. The decision is based on the exact parameters requested by the user and the corresponding resources available in the network. Thus, should a perpetrator attack the call processor of an ATM switch and successfully take over the control, the normal functioning of the ATM network may be greatly jeopardized.

6.3.1.6 *Trusting Traffic Controls at the UNI.* Following successful call establishment and negotiation of the traffic parameters between the user and the UNI, only the UNI checks the user traffic for compliance. In general, where user traffic exceeds the agreement, determined through a sliding window or other mechanisms, the excess cells are marked with 1 in their cell loss priority (CLP) field so that they may be dropped in the event of severe congestion within the network. The thinking is that the combination of the bandwidth allocation along the links and the traffic controls at the UNI will ensure the absence of congestion and undue cell loss. A key difficulty with the thinking is that ATM traffic is highly bursty, leading to uncertainty in the measurements of sustained cell rate and peak cell rates that are utilized in the negotiations. As a result, traffic control at the UNI is imprecise, permitting excess cells to slip into the network. This characteristic may be successfully exploited by a perpetrator, especially if access is gained to one of the intermediate ATM nodes. The problem is real, since for obvious performance reasons, compliance is not verified at the intermediate nodes, deep in the network.

6.3.1.7 *Access to Knowledge of the State of the Network.* A key ATM principle is that a node within a peer group possesses knowledge of the complete topology of the group, i.e. the location of the peer nodes and the links, plus the gateway node identifiers of other groups to which this peer group is connected. In addition, a node may contain updated information on the bandwidth utilization of links other than the ones to which it is directly connected, node processing delays at peer ATM switches, etc., which it utilizes to compute superior routes dynamically. The complete knowledge represents the current "state" of the network at the peer group level, from the perspective of the node in question. Clearly, where a perpetrator gains unauthorized access to an ATM switch, knowledge of the "state" may constitute a vulnerability.

6.3.2 A Methodology for Attack Modeling

Given the absence in the literature of a scientific methodology to derive attack models from analyzing network vulnerabilities, this section presents a new approach that has been developed and experimentally validated for ATM networks. The approach consists of three phases: designing an attack, modeling the attack and simulating it under realistic conditions, and assessing the performance impact of the attack model through appropriate metrics. The attack design phase, in turn, relies on two mechanisms. Under the first mechanism, a careful analysis of the key ATM principles and the modes in which the normal ATM functioning may be disrupted will most likely yield the straightforward vulnerabilities and the corresponding attacks. To uncover highly complex and subtle vulnerabilities with far-reaching impacts and sophisticated attacks, the second mechanism suggests that one or more individuals explicitly engage in intense reflection on the basic and fundamental nature of ATM networking and its constituent subprocesses. Reflection is defined as introspective contemplation or meditation on a thought or idea, quietly or calmly, with a view to understanding it in its right relation to all other concepts. Only after a thorough and deep analysis of all of the fundamental ATM issues has been completed, based on all that is known, and following an exhaustive and meticulous exploration of every available approach and reasoning technique, does the focus of the reflection converge. Given that networks will be progressively powerful, encompassing, useful, and complex, the second mechanism is likely to become indispensable to ensure their security.

To enhance the effectiveness of an attack model, the following guidelines may be employed:

- How to strengthen an attack's disruptive influence on normal ATM operation.

- How to combine the essence of different individual attacks into a complex attack.

- How to render an attack more difficult to detect.

The arguments underlying the use of an accurate, asynchronous, distributed simulator to emulate the attacks and study their performance behavior are as follows. First, a simulator is highly flexible and permits the testing of complex attacks that may be very difficult to realize in an actual ATM network. Second, the incorporation of hardware and timing parameters of the ATM switches into the simulator renders it highly realistic and implies accurate results. Third, modeling attacks on an actual ATM network may require extensive hardware and software modifications to the ATM switches, which may be prohibitively expensive.

To assess the effectiveness of the attacks, especially sophisticated attacks, existing metrics may require reexamination, or completely new metrics may need to be designed. The following guidelines have been employed in metric design:

- Specify the exact function of the attack.

- Identify network parameters upon which the impact of the attack may be observed.

- Quantify the extent of the adverse impact, relative to the normal operation.

- Develop reasoning about how the metric may be utilized to detect an attack.

6.3.3 Synthesizing Attacks

This section describes the synthesis of seven basic and two complex attacks, building on the vulnerabilities presented in subsection 6.3.1 and the methodology outlined above. The nature of basic and complex attack types was explained earlier in this chapter.

6.3.3.1 *Attack 1.*

6.3.3.1.1 *The Intent.*

Under this attack, an ATM switch under a perpetrator's control may choose to reject any incoming SETUP request, falsely citing lack of resource availability. The switch launches a REJECT message in the direction of the source node. This is a simple attack, its obvious intent being "denial of service." Under ATM principles, an intermediate ATM switch retains the right to accept or deny a call setup request, subject to the availability of VPI/VCI values, bandwidth, and other network resources. In theory, the ATM switch may reject every incoming call setup request. The DTL field in the SETUP message contains the source and destination node identifiers.

As a particular variation of this attack, the perpetrator may target specific nodes and choose to reject select incoming call requests, based on the following:

- Reject call requests that originate at the "nodes to be attacked."

- Reject call requests whose destination node identifiers are a subset of the "nodes to be attacked."

The rejected calls are labeled "target calls," and the rationale underlying this variation may be as follows. A node that rejects every call request through it will be an obvious suspect, and the perpetrator is likely to be detected quickly. By causing a limited number of specific call requests to be failed, the network will continue to exhibit normal behavior, and the perpetrator will be difficult to detect.

It is pointed out that only new call requests are impacted by this attack. Calls that are established prior to the onset of this attack are unaffected.

6.3.3.1.2 *Analysis of Attack Behavior and Design of Output Metrics:*

To analyze the impact of an attack, one needs to examine carefully the difference between the network's normal behavior and that under attack, both network-wide and at the node in question, say n. Since call rejection is the primary objective of the attack, the percentage of successfully established calls, network-wide as well as at n coupled with the number of rejected call requests at "n" will constitute key assessment metrics. The complete list of metrics for the normal and attack scenarios, is presented subsequently. It is pointed out that of the total number of unsuccessful calls, a few fail due to lack of resources, while the others have been subject to attack.

Normal Scenario:

- Number of call requests intercepted at each node: TN_n

- Number of successfully established calls (out of TN_n) at each node: SN_n

- Number of calls rejected (out of TN_n) at each node: RN_n

- Number of target calls (out of RN_n): GN_n

- Ratio of SN_n to TN_n, expressed as a percentage: PSN_n

- Ratio of GN_n to RN_n, expressed as a percentage: PGN_n

Under Attack Scenario:

- Number of call requests intercepted at each node: TA_n

- Number of successfully established calls (out of TA_n) at each node: SA_n

- Number of calls rejected (out of TA_n) at each node: RA_n

- Number of target calls (out of RA_n): GA_n

- Ratio of SA_n to TA_n, expressed as a percentage: PSA_n

- Ratio of GA_n to RA_n, expressed as a percentage: PGA_n

The analysis consists in comparing

- PSA_n and PSN_n, and

- PGA_n and PGN_n

6.3.3.2 Attack 2.

6.3.3.2.1 The Intent.

The intent of the attack is as follows. Consider that a call setup request, that originates at a "node to be attacked" arrives at an intermediate node, which is already under the perpetrator's control. Under the attack, the intermediate node accepts the connection and preempts further onward propagation of the SETUP message, with the result that the destination node never knows of the call request and then drops all traffic cells on the channel.

The goal of this attack is "indirect denial of service." The underlying rationale of the attack is as follows. If a call setup request is rejected, the user is likely to retry and may conceivably obtain a different route that bypasses the affected node. By creating a false impression that the call has been successfully established, the attack reaches a higher degree of deception. The behavior of the affected node under attack is explained in detail, as follows:

- The affected node waits until it receives a call SETUP request.

- The node does not execute the call acceptance control algorithm, and the SETUP message is not forwarded to the subsequent node, as specified in the DTL field.

- The node returns a CONNECT message toward the source node, creating a false impression that connection has been established.

- The SETUP message is fully contained at the affected node.

- Once the source node is assured that the call has been established successfully, it will initiate traffic cell propagation along the channel, all of which are discarded by the perpetrator when they reach the affected node.

As in attack 1, a variation of attack 2 may consist in the perpetrator targeting specific call requests to avoid detection. The attacker may focus on a list of target nodes, and any call request with the source or destination node contained in the list is subject to attack. Also, as in attack 1, here only new call requests are impacted by this attack. Calls that are established prior to the onset of this attack are unaffected.

A second variation of attack 2 may be conceived if, contrary to the practice of ATM Forum's PNNI 1.0 protocol, the source and destination addresses of the call request are retained at the intermediate ATM switches following successful call establishment. Under these circumstances, the attacker may compare the retained source and destination addresses of all active calls against its list of target nodes, and in the event of a match, the traffic cells of the corresponding call in the channel are dropped.

6.3.3.2.2 Analysis of Attack Behavior and Design of Output Metrics.

Analysis of the attack behavior must focus on the target calls and verify that they are correctly identified and successfully attacked. Given that the affected node does not forward the SETUP requests corresponding to the attacked calls, the bandwidth availability at the links beyond the affected node remains unaffected, and this is likely to impact network performance. The complete list of metrics for the normal and attack scenarios are presented subsequently.

Normal Scenario:

- Number of calls originating at or destined for the "node to be attacked": TN_n

- Successfully established calls (out of TN_n calls): SN_n

- Ratio of SN_n to TN_n, expressed as a percentage: PSN_n

- Average bandwidth utilization across all links in the network, measured at appropriate time intervals

Under Attack Scenario:

- Number of calls originating at or destined for the "node to be attacked": TA_n

- Successfully established calls (out of TA_n) calls): SA_n

- Ratio of SA_n to TA_n, expressed as a percentage: PSA_n

- Average bandwidth utilization across all links in the network, measured at the same time interval as in the normal scenario.

The analysis consists in comparing PSN_n with PSA_n and the average bandwidth utilization across all links between the normal and attack scenarios.

6.3.3.3 Attack 3.

6.3.3.3.1 Intent.

Under this attack, a perpetrator controlling an intermediate node behaves as follows. While a call SETUP request originating at a node contained in the list of target nodes is in progress, the perpetrator refrains from any interference. In the event that the call is successfully established, subsequent traffic cells sent from the originating node are dropped at the affected node. The rationale underlying attack 3 is as follows. Unlike under attack 2 as detailed in subsection 6.3.3.2, where the increased bandwidth availability at the links beyond the affected node may manifest in effects that may reveal the perpetrator, attack 3 limits its influence on the bandwidth allocation. Thus, from the perspective of an attacker focused on minimizing detection, attack 3 constitutes an improvement over attack 2. The focus of attack 3 is to cause traffic loss, which may force the higher layers to repeatedly retransmit traffic, thereby increasing network congestion and further disruption in normal network operation.

As further refinement of attack 3, consider the following argument. Given that the affected node will drop traffic cells, according to plan, the subsequent intermediate nodes will fail to receive the expected traffic and may report an anomaly to the network administrator. To further minimize detection, the perpetrator may adopt one of two approaches:

- Drop only a few cells, at random.

- Drop all the traffic cells on the channel, periodically, for short bursts of time intervals. The attack can be highly effective where the perpetrator is an insider and is privy to the exact sensitivity of the traffic.

6.3.3.3.2   *Analysis of Attack Behavior and Design of Output Metrics.*

Logically, the best metrics to observe manifestations of this attack are the cell drop rate and the time variation of the cell drop, at every channel of the switch fabric network at each node of the network. The complete list of metrics for the normal and attack scenarios are as follows. The symbol c reflects the channel.
   Normal Scenario:

- Number of cells transmitted by the source node on a channel: SN_c

- Number of cells received at the destination over the same channel: RN_c

- Number of cells dropped on the channel: SN_c - RN_c

Under Attack Scenario:

- Number of cells transmitted by the source on a channel: SA_c

- Number of cells received at the destination over the same channel: RA_c

- Number of cells dropped on the channel: SA_c - RA_c

The analysis consists in comparing the quantities $(SN_c$ - $RN_c)$ and $(SA_c$ - $RA_c)$.

6.3.3.4   *Attack 4.*

6.3.3.4.1   *Intent.*

Given that the traffic controls are located only at the UNIs, a perpetrator may easily induce disruption into the network by synthesizing an arbitrary number of bogus ATM cells, intermixing them with authentic cells and then transmitting them along the established channels. The ATM switches check neither for the content of the traffic cells nor for their rate relative to the negotiated contract. As a result, while bandwidth may be overutilized, resulting in lost legitimate cells, the end systems may upon reassembly either misinterpret the information contained in the bogus cells or conclude the transmission as corrupted and request retransmission.
   Given that multiple channels may be active at any time within a switch, the attacker may resort to either of the following choices:

- Select, arbitrarily, either one or a subset of all currently active channels.

- Identify channels where the source and destination nodes are contained in the attacker's target list.

6.3.3.4.2 *Analysis of Attack Behavior and Design of Output Metrics.*

The complete list of metrics for the normal and attack scenarios, is as follows. The symbol c represents the channel identifier.
Normal Scenario:

- Number of cells transmitted by the source node on a channel: SN_c

- Number of cells received at the destination node along the same channel: RN_c

- Number of cells dropped in the channel: SN_c - RN_c

- Number of cells received at a node: RN_n

- Number of cells forwarded by the node: SN_n

- Number of cells dropped at the node: RN_n - SN_n

Under Attack Scenario:

- Number of cells transmitted by the source node on a channel: SA_c

- Number of cells received at the destination node along the same channel: RA_c

- Number of cells dropped in the channel: SA_c - RA_c

- Number of cells received at a node: RA_n

- Number of cells forwarded by the node: SA_n

- Number of cells dropped at each node: RA_n - SA_n

The analysis consists in the following comparisons:

- $(SN_c$ - $RN_c)$ and $(SA_c$ - $RA_c)$

- $(RN_n$ - $SN_n)$ and $(RA_n$ - $SA_n)$

6.3.3.5 *Attack 5.*

6.3.3.5.1 *The Intent.*

Under this attack, for a call already established between two given nodes in the network, a perpetrator occupying an intermediate node will first examine whether the source or destination node for the call occurs in its list of target nodes. If affirmative, all traffic cells are diverted to other active channels, selected periodically and at random. The design of attack 5 combines the essence of attacks 3 (subsection 6.3.3.3) and 4 (subsection 6.3.3.4) while attempting to increase resistance to detection. Thus, instead of completely dropping all traffic cells as in attack 3, or synthesizing and propagating bogus cells as in attack 4, here the cells are diverted from the legitimate channel onto other active channels.

Under this attack the disruptions include the following. First, bandwidth is consumed illegally, causing legitimate cells to be dropped. Second, the presence of excess cells along an unintended channel will increase the processing overhead at the subsequent ATM switch and possibly cause buffers to overflow. Third, the recipient at the destination node will attribute the absence of the expected traffic to congestion, and the higher-layer protocols may request retransmission, thereby contributing to congestion. Fourth, the presence of unexpected cells at the nodes may lead the recipients to wrongly conclude that the traffic has been corrupted and request retransmission, further aggravating the congestion problem.

A key characteristic of the attack is that it is launched only after the appropriate call has been successfully established. Its behavior is expressed as follows:

- The affected node contains the perpetrator's list of target nodes.

- When a SETUP message arrives at the affected node, the perpetrator examines whether the source or destination node is included in the list of target nodes.

- If affirmative, the attacker waits until the call is established successfully.

- If the call request fails for any reason, the attack no longer applies.

- If the call is established successfully, label it as the "target call," and the following procedure is initiated:

    - In the VPI/VCI mapping table, update the entry for this channel as follows.

    - Select an output VPI/VCI pair from any of the already existing channels arbitrarily and write it as the output VPI/VCI for this channel.

    - Continue to modify the output VPI/VCI pair entry for this call periodically, randomly choosing a VPI/VCI pair from the mapping table that reflects the current, active channels.

    - When a call is about to be terminated, the perpetrator examines whether it is currently utilizing the corresponding VPI/VCI pair. If affirmative, the perpetrator will utilize a different VPI/VCI pair from the list of active channels. If negative, no action is needed.

    - While the entry for the target call is being updated, if the list of available VPI/VCI pairs is nil, the cells are dropped.

6.3.3.5.2 *Analysis of Attack Behavior and Design of Output Metrics.*

The complete list of metrics for the normal and attack scenarios, is as follows. Comparison of the corresponding metric values is expected to yield insights into the attack behavior:

- Bandwidth allocation at each link.

- Bandwidth utilization at each link.

- Number of cells transmitted by the source nodes for each call.

- Number of cells received at the destination nodes, for each call.

- Cell drop rate at each node, in percentage, as a function of the total number of cells received at the node.

6.3.3.6 *Attack 6.*

6.3.3.6.1 *The Intent.*

The intent of this attack is to reduce the available bandwidth and VPI/VCI values in the network by creating bogus call requests with the affected node as the origin and other nodes as the destinations. Clearly, legitimate call requests in the network are likely to be adversely affected from lack of adequate network resources. Given that the current UNI 3.0 standard does not preclude a user from initiating multiple independent call requests, either to a single destination node or different destination nodes, attack 6 may be easily launched at the user-level, and it requires little knowledge of the ATM network operation. When launched from the user-level, this attack requires the perpetrator to possess knowledge of other node identifiers that may serve as potential destination nodes. A key limitation of the attack, however, is that the perpetrator cannot influence the synthesis of the DTL field. That is, the perpetrator cannot insist on the choice of the intermediate nodes in the route computation.

Attack 6 may be exploited to cause extensive disruption in the event the perpetrator controls a node and has access to the topology of the group, i.e., the location of the links and its peer nodes. If the affected node serves as a gateway node, it knows the addresses of the adjacent nodes in other groups. Since the DTL field is synthesized by the call processor in an ATM switch, the perpetrator will very likely create bogus calls with carefully considered routes to infect a maximal number of links and nodes in the network.

As an example, consider the topology shown in Figure 6.11. Assuming that the attacker is located at node 2, it may establish bogus calls with its peer nodes, specifying the exact routes as shown subsequently, with the result that every link and node in the group is affected.

- node 2 – node 1 – node 6 – node 5

- node 2 – node 3 – node 6

- node 2 – node 4 – node 5

Relative to the implementation of this attack, the attacker has two choices, described below under cases A and B.

- **Case A:** The perpetrator may originate a few bogus call requests, each specifying a substantial bandwidth demand. While the lack of adequate bandwidth along the links will clearly cause genuine user calls to be rejected, the large bandwidth request is likely to be flagged by the network administrator as suspicious.

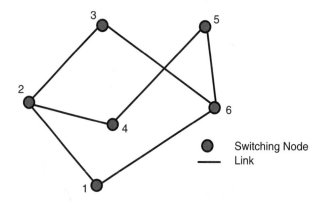

Figure 6.11 Understanding the nature of attack 6.

- **Case B:** The perpetrator may initiate a large number of call requests, each specifying a relatively small bandwidth demand. While such calls would appear normal and defy detection, their cumulative bandwidth occupancy will have the same devastating effect as in case A. Furthermore, with the bandwidth request being low, it is likely that fewer bogus call requests will be rejected than in case A.

6.3.3.6.2 *Analysis of Attack Behavior and Design of Output Metrics.*

Clearly, bandwidth availability at the links and call success rate constitute the key metrics for this attack. The complete list of metrics for the normal and attack scenarios is as follows:

- Bandwidth availability at each link (l): BW_l

- Number of calls initiated at each node (n): OC_n

- Number of successful calls at node n (out of OC_n): SC_n

- Ratio of successful calls to total number of initiated calls at each node (n), expressed as a percentage: PC_n

6.3.3.7 *Attack 7.*

6.3.3.7.1 *The Intent.*

The conception of attack 7 stems from the absence of explicit safeguards against tampering with the key elements of the ATM switch fabric including the buffers. Under this attack, a perpetrator occupying an affected node may artificially reduce the size of the buffer available to serve the ATM traffic, thereby causing increased cell loss and possibly affecting the cell delay characteristics. Where the perpetrator lacks explicit knowledge and control over the buffer space, cell loss may be accelerated by synthesizing bogus cells as described under attack 4 in subsection 6.3.3.4.

6.3.3.7.2 *Analysis of Attack Behavior and Design of Output Metrics.*

Metrics similar to those utilized for attacks 3 and 4, explained in subsections 6.3.3.3 and 6.3.3.4, may be utilized to study the manifestations of this attack. The complete list of metrics for the normal and attack scenarios, is as follows. The symbol c represents the channel identifier. Normal Scenario:

- Number of cells transmitted by the source node on a channel: SN_c

- Number of cells received at the destination node along the same channel: RN_c

- Number of cells dropped in the channel: SN_c - RN_c

- Number of cells received at a node: RN_n

- Number of cells forwarded by the node: SN_n

- Number of cells dropped at the node: RN_n - SN_n

- Delay incurred by traffic cells on each channel (c): DN_c

Under Attack Scenario:

- Number of cells transmitted by the source node on a channel: SA_c

- Number of cells received at the destination node along the same channel: RA_c

- Number of cells dropped in the channel: SA_c - RA_c

- Number of cells received at a node: RA_n

- Number of cells forwarded by the node: SA_n

- Number of cells dropped at each node: RA_n - SA_n

- Delay incurred by traffic cells on each channel (c): DA_c

The analysis consists in the following comparisons:

- $(SN_c$ - $RN_c)$ and $(SA_c$ - $RA_c)$

- $(RN_n$ - $SN_n)$ and $(RA_n$ - $SA_n)$

- DN_c and DA_c

6.3.3.8 *Attack 8: Complex Attack.*

6.3.3.8.1 *The Intent.*

Unlike basic attacks 1 through 7, attacks 8 and 9, described subsequently, are complex. They are based on a malicious intent or objective, and are termed focused attacks. Although their effects are generally severe, these attacks are not always guaranteed to be 100% successful. A key focus of these attacks is that they tend to split the network by isolating nodes and disrupting the links, i.e., rendering them virtually unusable. A few relevant definitions are the following:

- **Severity:** A measure of severity of a complex attack, launched by an attacker occupying a node on the network at its own peer group level is the degree to which it successfully disconnects the individual nodes. If the disruption, either complete or partial, of a link causes N distinct pairs of nodes to be disconnected, then the severity is N. The larger the value of N, the more extensive the damage.

- **Sensitive Node:** A sensitive node is one that when isolated from the remainder of the network can bring about maximal disruption to the network connectivity.

- **Sensitive Link:** A sensitive link, similar to a sensitive node, is one that when brought down can cause maximal disruption to the network connectivity.

The intent of attack 8 is to isolate a particular target node, say T, in the network at the current peer group level and completely deny service to and from it. Assume that the attacker controls node A in the network at the peer group level. Figure 6.12 shows both nodes T and A. Clearly, the attack plan is for the perpetrator to synthesize bogus calls intelligently and consume the maximal resources held by T. The principal resources include the VPI/VCI values associated with the links connected to T, buffer space at T, and the available bandwidth on the links going

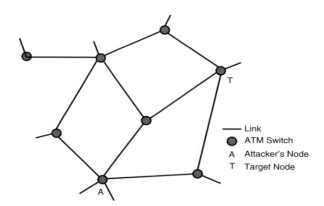

Figure 6.12 Complex attack 8.

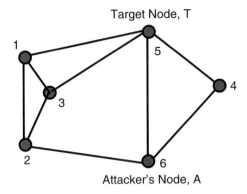

Figure 6.13 Design of complex attack 8: nodes T and A are directly connected.

out from the node T. To consume these resources, the attacker may establish bogus calls from A to all of the neighbors of T, choosing routes through T. For maximal impact, the bogus calls may be synthesized by first analyzing the topology of the network.

The design of attack 8 for different topologies is explained subsequently.

In the topology in Figure 6.13, the attacker node A and target node T are connected directly. Under this scenario, the perpetrator may attempt to establish bogus connections originating at A, with all of the neighbors of T, choosing routes through T such that the bandwidth of the links to and from T are maximally consumed. The attacker may choose to establish either a few calls, each with significant bandwidth demand, or many calls, each with modest bandwidth allocation request.

In the topology in Figure 6.14, the attacker node A and target node T are connected indirectly. As in the previous scenario, the perpetrator may attempt

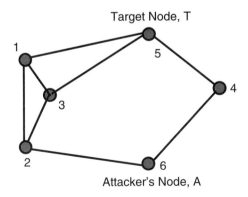

Figure 6.14 Design of complex attack 8: nodes T and A are connected indirectly.

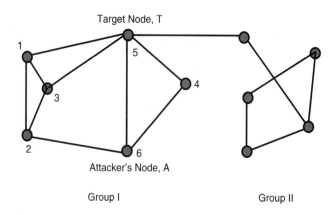

Figure 6.15 Design of complex attack 8: T is a gateway node.

to establish bogus connections originating at A, with all of the neighbors of T, choosing routes through T such that the bandwidth of the links to and from T are maximally consumed. The attacker may choose either to establish a few calls, each with significant bandwidth demand, or many calls, each with modest bandwidth allocation request. Thus, the routes for the bogus call requests may assume the forms A-4-5-3, A-4-5-1, A-2-3-5-4, and A-2-1-5-4.

In the topology in Figure 6.15, the target node T is a gateway node. In addition to adopting the same approach as in the previous two scenarios, the attacker may attempt to establish a connection to the gateway node in group II, routed through T. This effort is helped by the PNNI protocol, where the peer group leader disseminates all information to every node of the peer group including the identifiers of the inter-group links and the gateway nodes in other groups. However, under this attack, A cannot affect service to target node T from group II.

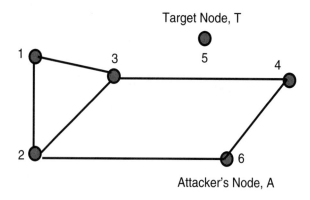

Figure 6.16 Design of complex attack 8: T is an isolated node.

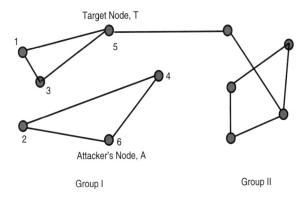

Figure 6.17 Design of complex attack 8: T and A in disconnected subgroups of a peer group.

In the topology in Figure 6.16, the target node T is isolated from the remainder of the network. Although, strictly, this is an irrelevant case in that the existence of T will be unknown to A, the situation may be the result of a loss of connection between T and the remainder of the group.

In the topology in Figure 6.17, T and A are elements of two distinct disconnected subgroups of a peer group. As with the previous scenario, this is also irrelevant, since by definition, every member of a group must be aware of the existence of other current members of the peer group. This scenario, however, serves as an inspiration to conceiving another severe attack, discussed subsequently in subsection 6.3.3.9.

6.3.3.8.2 *Analysis of Attack Behavior and Design of Output Metrics.*

Since the measure of severity of attack 8 is defined by the degree to which an individual node is disconnected from the remainder of the network, the impact of attack 8 may be encapsulated primarily through the call rejection rate. The complete set of metrics for both normal and attack scenarios is presented as follows:

- Number of call requests initiated at each node.

- Number of call requests rejected that were initiated at the current node.

- Cumulative number of call requests initiated by all nodes of the network.

- Cumulative number of call requests rejected in the entire network.

- Cumulative number of initiated call requests with T in their DTL fields: "TargetedCalls"

- Cumulative number of successful calls (out of "TargetedCalls"): "STargetedCalls"

6.3.3.9 *Attack 9: Complex Attack.*

6.3.3.9.1 *The Intent.*

The objective of this attack is to disrupt communication along a specific link, thereby affecting the two nodes that are directly connected by the link. Evidently, through an intelligent analysis of a peer group's topology, a perpetrator may focus on a critical link which when brought down will severely affect connectivity and reachability within the group. In the worst case, the link may be so sensitive that its disruption may cause the peer group to be split into two isolated subgroups.

In Figure 6.18, under attack 9, the attacker at node A targets the link connecting nodes N1 and N2. First, if a call between N1 and N2 is routed through A, the perpetrator can trap the call SETUP request message and send a reject as described under attack 1 in subsection 6.3.3.1. In addition, the attacker may employ attacks 2 and 3 as described in subsections 6.3.3.2 and 6.3.3.3, respectively. Second, when calls between N1 and N2 are not routed through A, the perpetrator may employ any of the following approaches:

- The perpetrator initiates bogus calls to node N2, choosing as many different routes as possible, all of which utilize N1 as an intermediate node. The perpetrator also initiates bogus calls to node N1, specifying as many different paths as possible, all of which are routed through N2. For each of these bogus calls, maximal bandwidths are requested to cause the most disruption.

- The perpetrator may either flood the network with database summary packets that indicate that nodes N1 and N2 lack adequate resources or deliberately introduce erroneous hello packets that indicate that the connection between nodes N1 and N2 is broken. As a result, in computing routes for call requests, other nodes in the network will avoid including nodes N1 and N2 and therefore the link between N1 and N2. A weakness of this attack is that the disruption it causes may last only for a short time. Even though

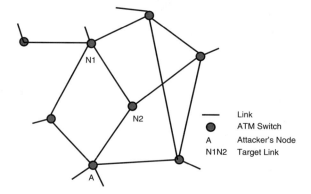

Figure 6.18 Complex attack 9.

a temporary inconsistency may be introduced into the topology database, it would be difficult for the perpetrator to calculate and predict the extent of the damage. Database summary packets and hello packets are sent continuously by all the nodes in the network, and the deliberately introduced erroneous packets may be overwritten and discarded.

Under attack 9, connections that do not utilize the link between N1 and N2 but choose N1 and N2 as intermediate nodes may also be affected.

6.3.3.9.2 *Analysis of Attack Behavior and Design of Output Metrics.*

The complete set of metrics for both normal and attack scenarios is similar to that of attack 1 described in subsection 6.3.3.8, and is presented as follows:

- Number of call requests initiated at each node.

- For every node, number of originating call requests that are eventually rejected.

- Cumulative number of call requests initiated by all nodes of the network.

- Cumulative number of call requests rejected in the entire network.

- Cumulative number of call requests in the network that specify link (N1 ↔ N2) in their DTL fields: "TargetedCalls"

- Cumulative number of successful calls (out of "TargetedCalls"): "STargetedCalls"

6.4 Modeling, Simulation, and Behavior Analysis of Security Attack Models

The objectives of modeling and simulation of an attack model are twofold. First, they serve to verify that the attack model design successfully forces the corresponding vulnerability to be manifested at an observable output, under realistic input traffic conditions. Second, they are expected to yield an accurate picture of an attack's qualitative or quantitative impact on one or more of the network's performance metrics. To achieve these objectives, three requirements are in order. The choice of a representative ATM network topology satisfies the first requirement. The second requirement is the development of an input traffic stimulus that is representative of the real world. The third requirement is the development of ATMSIM 1.0 [94], an accurate, distributed ATM simulator that utilizes an asynchronous distributed simulation algorithm to execute the computationally intensive behavior models of the ATM switches as well as emulate the complex asynchronous interactions between them. In this section, five attack models, representative of the types of attacks described earlier are modeled and simulated. The design of attack models starting from vulnerability analysis, coupled with their modeling, simulation, and behavior analysis, constitutes a scientific approach to studying security attacks in ATM networks.

6.4.1 Choice and Justification of the Network Topology

A 9-node representative ATM network topology is synthesized for this study and shown in Figure 6.19. Given that hundreds of simulation runs are required for debugging and performance analysis, simulation is computationally intensive, and since only finite computational resources are available, the choice of a reasonably sized network topology with a modest number of nodes is necessary to complete the study in a reasonable time. To exercise the key elements of the PNNI protocol and to achieve acceptable accuracy in the overall performance analysis, the network must consist of multiple peer groups. In this study, a total of 9 nodes are organized into three peer groups, arranged in a two-level hierarchy. The groups are labeled A, B, and C. Of these peer groups, one in particular, group B, is designed to consist of 5 nodes, more than in groups A and C. This enables peer group B to offer greater flexibility in the placement of the attacker node versus the target nodes and also to model the more complex attack 5. Peer groups A and C consist of two nodes each and while both nodes of C are of gateway type, i.e., they interface with other peer groups, only one node of A is of gateway type. The aim underlying the node placements and their connectivity is to yield insights into the overall network behavior under attack.

The two nodes in peer group A are labeled A.1 or 1 or 100 and A.2 or 2 or 101. In peer group C, the nodes are labeled C.1 or 8 or 120 and C.1 or 9 or 121. In peer group B, the nodes are labeled B.1 or 3 or 110, B.2 or 4 or 111, B.3 or 5 or 112, B.4 or 6 or 113, and B.5 or 7 or 114. The multiple labels underlie different naming conventions used by ATMSIM 1.0 and other associated programs. The network topology is conceived from the following interconnected cities in the US. Nodes A.1, A.2, B.1, B.2, B.3, B.4, B.5, C.1, and C.2 correspond to Olympia, Carson City, Pierre, Jefferson City, Baton Rouge, Topeka, Santa Fe, Rayleigh, and Albany. Links are bidirectional, and separate identifiers are used to label each direction. Thus, $4 \rightarrow 9$ represents a link from node 4 to node 9. Table 6.1 presents

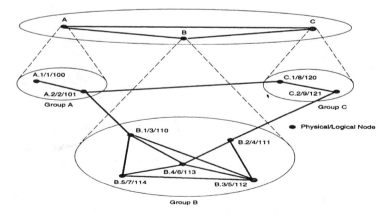

Figure 6.19 9-node representative ATM network topology.

Link x↔y	Link ID x → y	Link ID x←y
A1↔A2	1	14
A2↔B1	2	15
B1↔B3	3	16
B1↔B4	4	17
B1↔B5	5	18
B2↔B3	6	19
B2↔B4	7	20
B3↔B4	8	21
B3↔B5	9	22
B4↔B5	10	23
B2↔C2	11	24
C1↔C2	12	25
A2↔C1	13	26

Table 6.1 Identifiers for each of the bidirectional links in the network.

the identifiers for each of the links in the network and Table 6.2 presents their key characteristics, namely, their distances and physical propagation delays in absolute seconds and in terms of timesteps, the basic unit of time in the simulation. Assuming OC-3 links, rated at 155.52 Mb/s, and an ATM packet size of 53 bytes, or 424 bits, the transmission time for an ATM packet is $(424b)/(155.52 \text{ Mb/s}) = 2.73$ μs.

Link	Distance (D) in km	D km/(Fiber Optic Speed = 200,000 km/s) in sec	Link Delay in Timesteps
A1-A2	1158.16	5.79	2121
A2-C1	4420.08	22.1	8095
A2-B1	2369.90	11.84	4337
C1-C2	1022.04	5.11	1872
C2-B2	1546.73	7.73	2832
B1-B3	2405.78	12.03	4407
B1-B4	927.91	4.64	1700
B1-B5	1486.07	7.43	2722
B2-B3	1268.37	6.34	2322
B2-B4	356.23	1.78	652
B3-B4	1505.06	7.5	2747
B3-B5	1916.00	9.58	3509
B4-B5	1347.38	6.74	2469

Table 6.2 Distances and physical propagation delays of the links.

6.4.2 Utilizing an Accurate ATM Simulator: ATMSIM 1.0

ATMSIM 1.0 [94] is an asynchronous distributed ATM simulator that encapsulates the key elements of ATM Forum's PNNI 1.0 and UNI 3.0. It utilizes the principles of asynchronous distributed simulation outlined in [93]. In this study, ATMSIM 1.0 is executed on a testbed of Pentium laptops, configured under the

Linux operating system and interconnected through a 100 Mb/s Fast Ethernet. While ATMSIM 1.0 incorporates input traffic [105], designed to emulate a representative ATM network, it permits users to generate call SETUP requests externally and insert them into the network. Traffic consists of audio, video, and data. While call SETUP requests are assumed to obey a Poisson distribution, the traffic volume is set based on the "stability criterion" [93]. The bandwidth requests are assumed to be stochastic, varying between 4 and 6 Mb/s for normal calls. For bogus calls, the bandwidth requests are defined by the specific attack and are discussed subsequently. In the course of its execution, the simulator writes intermediate data onto specific files, from which the progress of an attack may be inferred. For performance analysis, a total of 150 simulation runs are executed on the testbed, each run requiring approximately 10 12 hours of wall clock time, and cumulatively generating a total of 175 Mb of simulation data.

6.4.2.1   *Input Files.*

ATMSIM 1.0 permits the simulation to be controlled through a number of input files, the names and contents of which are described subsequently.

Files with name "atm_input_#": There are nine files, one for each node, represented by #, through which externally generated call SETUP requests may be inserted into the simulation. In addition, for successfully established external calls, these files serve as conduits for externally generated traffic cells to be inserted into the simulation. Information in these files is organized as lines with following fields: "timestep," "call_id," "action-character":

The "timestep" field specifies when this input line should be called into effect; the "call_id" field specifies a unique call identifier, and the "action-character" field contains one of two symbols, C or T. The symbol C represents a call SETUP request, and the data in subsequent lines are interpreted according to the following fields: "dest_id," "qos_class," "cpr," "ctd," where "dest_id" implies the destination node identifier, "qos_class" represents the quality of service class, not currently used, "cpr" refers to the peak cell rate, and "ctd" represents the cell transfer delay, also currently unused.

The symbol T represents a request to send user traffic, and the data in subsequent lines consist of a single field, "# cells," which refers to the number of cells to be inserted at "timestep" on behalf of the call "call_id," where it is successfully established. In the event the call fails, the information in the "# cells" field is ignored.

File with name "attack.in": Through this file the user specifies the attack to be activated. The information in the file is organized through lines with the fields: "attack#," "attacker," "target," where the first field, "attack#", identifies the specific attack by number; the second field, "attacker", specifies the node identifier that is controlled by the perpetrator; and the third field, "target," specifies the target node identifier, i.e., the node to be attacked. The node identifiers are specified utilizing the three-digit naming convention explained earlier in subsection 6.4.1.

File with name "ATM_PARAMS": The information in this file is organized through three lines, shown subsequently. The first and second lines specify the input and output buffer sizes at each node, in terms of ATM cells. The information in the third line is relevant to attack 2, elaborated subsequently in subsection 6.4.4, and it specifies the time period, i.e., inverse of the frequency, with which different erroneous destination nodes must be selected.

IN_BUF_SIZE 30000
OUT_BUF_SIZE 30000
ATTACK_2_FREQ 25000

6.4.2.2 *Output Files.* The simulator logs data into the the output files, which, in turn, serve as a window into the internal functioning of the simulation, under normal and attack scenarios.

Files with name "atm_log_file_#": There are 9 such files, one per node, represented by #, wherein each node dynamically logs key information. The information contained is of three types:

- [110000] **CALL ID:1 SETUP:UNI SETUP:**
  calling party: UNIie_CGP_NUM:113 :0:0
  called party: UNIie_CP_NUM:101 :0
  ATD:ATM_TRAF_DESC: for_pcr_0 (130/1000)
  conn id: flags:0 vpci:0 vci:0

  The implication here is that at a time instant specified by **110000** timesteps, the call SETUP request with identifier **1** was initiated at this node, with the destination node 101. The QoS parameters requested are identified in the field "ATM_TRAF_DESC for_pcr_0 (130/1000)" consisting of a bandwidth request of **1000**.

- [110217] **CTIME CALL ID:1**
  [110217] **CALL SETUP SUCCESSFUL:DTL**$[trans\_0][101|0][110|0]$

  The information contained here is that a call request with identifier **1** was successfully established at a time instant given by **110217** timesteps. The DTL for this call is contained at the end of the line.

- [110000] **BW_AVAIL: link [3|6]:155000**

  This indicates at the time instant given by 110000 timesteps, the available bandwidth on the link between nodes 3 and 6 is 155000. This is reserved bandwidth, not actual, and appears in the log file only when there is a change.

Files with name "atm_output_#": These files, one for each node, contain the summary statistics for the corresponding node. The summary includes the number of ATM cells processed, number of cells dropped, etc.

6.4.3   Attack 1

In essence, this attack constitutes a realization of basic attack 2, previously described in subsection 6.3.3.2. It is restated as follows:

If a call setup request that originates at a "node to be attacked" arrives at an intermediate node (which is already under attacker's control), under the attack condition, accept the connection at this intermediate node without letting the destination know about it and then drop all the cells on that channel.

Assume that the perpetrator controls node B.2 and its target is node B.5. Ideally, the perpetrator should occupy a node from where it may intercept the maximal number of target calls. In the network in Figure 6.19, nearly all of the nodes are connected to each other, and no node has any significant advantage over another. Since the gateway nodes B.1 and B.2 are connected to other groups, they offer an ideal location to a perpetrator to inflict maximal damage to the overall network. Both B.1 and B.2 offer similar connectivity, and the latter is chosen arbitrarily. As a challenge to the attacker, the target node is located as far away from B.2 as possible. Both nodes B.1 and B.5 are strong candidates, and B.5 is selected arbitrarily. Thus, all call SETUP requests originating at B.5 and routed through B.2 such as to destination nodes C.1 or C.2 are likely to be intercepted by the attacker. Thus, the input file "attack.in" will contain the line: "1 111 114."

6.4.3.1 *Experiments.* Observations indicate that of the approximately 550 call SETUP requests generated by the traffic generator incorporated within ATMSIM 1.0, only a handful, namely 6, originate at B.5 and include B.2 in their route. Thus, a number of trial experiments were designed, executed, and critically analyzed with the intent of identifying a scenario, where possible, that clearly reveals the impact of the attack. The initial experiments indicate that for externally generated calls, orignating at B.5 and destined for a node in peer group C, the route is more than likely to include the gateway node B.2 as an intermediate node. These calls are termed target calls.

To understand the relationship between the increasing intensity of attack and network performance, an experiment was designed where the number of externally generated target calls is gradually increased in steps from 8, to 17, 25 and 60. Thus, the input file "atm_input_7" is modified to include more calls with a node in peer group C as the destination.

6.4.3.2 *Analysis of the Simulation Results.* For every experiment a number of simulation runs are executed, and the data obtained from the output log files "atm_log_file_#" are processed to yield statistics on the available bandwidth at each link at different timesteps, number of successful calls at each node, and number of call failures. Table 6.3 presents the call success rates under normal and attack scenarios for eight externally inserted target calls. For the network, refer to Figure 6.19.

In Table 6.3 the call success rates at nodes A.1 and A.2 are unchanged under the attack scenario. This is logical, since they do not constitute target nodes. For node B.5, the call success rate is appreciably lower in the attack scenario, implying that the perpetrator at B.2 is successfully intercepting calls originating

Node	A.1	A.2	B.1	B.2	B.3	B.4	B.5	C.1	C.2
Call Success Rate (normal) %	35	53	59	40	63	63	62	43	38
Call Success Rate (attack) %	35	53	59	42	65	62	55	43	40

Table 6.3 Call success rates under attack 1 for 8 target calls.

at B.5. However, while the cause of the very slight decrease in the call success rate at node B.4, from 63% to 62%, is unclear, since B.4 is not a target node, the increase in the call success rates at nodes B.2, B.3, and C.2 is unexpected.

To gain insights into the unexpected behavior of the network under attack, the bandwidth availability at all of the links, under normal and attack scenarios, is critically examined. Three distinct behaviors are observed, two of which are represented in Figure 6.20 in the form of comparative analyses of the bandwidth availability at two links, $4 \rightarrow 9$ and $5 \rightarrow 4$, as a function of the progress of simulation. The third behavior is one where a link's available bandwidth is unaffected by the attack, implying that the link is beyond the sphere of influence of the attack. An example of the third behavior consists of the link $1 \rightarrow 2$. In Figure 6.20 it is observed from the graphs corresponding to the link $4 \rightarrow 9$ that more bandwidth is available while the attack is in progress. This is logical, since when node 4 or B.2 intercepts the target calls destined for peer group C, either node C.1 or node C.2, it neither forwards the call requests nor allocates bandwidth for these calls along the link $4 \rightarrow 9$. Similar behavior is observed for the link $9 \rightarrow 8$, but that is not shown here graphically.

In Figure 6.20, for the link $5 \rightarrow 4$, less bandwidth is available while the attack is in progress. A possible reason underlying this behavior is as follows. A number of calls originating at B.3 and destined for C.2 through B.2 may have failed under the normal scenario, due to lack of bandwidth availability along the link $4 \rightarrow 9$. This implies greater bandwidth availability along link $5 \rightarrow 4$. Under attack scenario, with less competition from the target calls, i.e., increased bandwidth availability along link $4 \rightarrow 9$, these previously failed calls are now successfully established, thereby reducing the bandwidth availability along link $5 \rightarrow 4$. Similar behavior is observed for the links $7 \rightarrow 6$ and $8 \rightarrow 2$.

The insight thus gained may help in understanding the unexpected increase in the call success rates at nodes B.2, B.3, and C.2, noted in Table 6.3. The increased available bandwidth along link $4 \rightarrow 9$ under the attack scenario permits calls originating at B.2 and B.3 and destined for C.2 or C.1 to succeed that may have failed under the normal scenario due to lack of bandwidth. As a result, the call success rates at B.2 and B.3 are higher under the attack scenario. Similarly, the increased available bandwidth along link $9 \rightarrow 8$ contributes to a higher call success rate at node C.2.

A series of experiments was designed for an increased number of target calls, extending from 17 to 25 to 60. Preliminary experiments reveal that beyond 60 target calls, the general behavior of the network under attack remains unchanged, although the effects are amplified. Table 6.4 presents the call success rates for each of the nine nodes, under normal and attack scenarios, while Figure 6.21 plots the

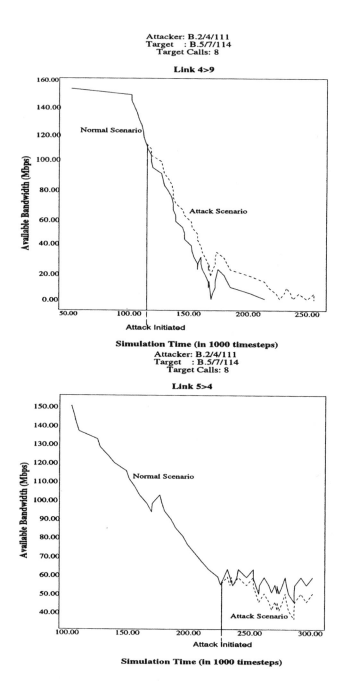

Figure 6.20 Bandwidth analysis under attack 1 for 8 target calls.

difference in the available bandwidths between the normal and attack scenarios, averaged over all the nodes, as a function of the number of target calls, for the representative links 4 → 9 and 5 → 4.

Analysis of the information in Table 6.4 and Figure 6.21 reveals that the behaviors previously observed for the eight target calls remain in effect and are amplified for higher numbers of target calls. Figure 6.21 further shows that while the difference in the available bandwidths between the normal and attack scenarios, say ΔBW, increases with increasing number of target calls, the rate of increase of ΔBW slows down at higher numbers of target calls, possibly due to the links being saturated with successful calls.

In summary, insights into the behavior of the network under this attack may be expressed as follows. Under normal input traffic and stable network operation, call success rates at the nodes are generally held at uniform values, subject to minor variation. Any sharp variation in the call success rate at one or more of the nodes is more than likely to indicate the onset of this attack. In general, under attack, links leading away from the node under the perpetrator's control are likely to experience higher bandwidth availability. In contrast, links leading into the node under the attacker's control may experience, depending on the traffic, higher bandwidth usage, i.e., lower bandwidth availability. Although there are exceptions such as links 9 → 8, 8 → 2, and 7 → 6, the probability is high that the link bandwidth usage around the node under the attacker's control will reveal a change under the normal and attack scenarios. This may constitute a valuable tool in localizing the source of the attack.

Target Calls: 8									
Node	A.1	A.2	B.1	B.2	B.3	B.4	B.5	C.1	C.2
Call Success Rate (normal) %	35	53	59	40	63	63	62	43	38
Call Success Rate (attack) %	35	53	59	42	65	62	55	43	40

Target Calls: 17									
Node	A.1	A.2	B.1	B.2	B.3	B.4	B.5	C.1	C.2
Call Success Rate (normal) %	35	53	58	40	61	62	60	43	37
Call Success Rate (attack) %	35	53	58	42	66	61	50	43	40

Target Calls: 25									
Node	A.1	A.2	B.1	B.2	B.3	B.4	B.5	C.1	C.2
Call Success Rate (normal) %	35	53	58	39	60	60	57	43	36
Call Success Rate (attack) %	35	53	57	43	67	60	47	43	40

Target Calls: 60									
Node	A.1	A.2	B.1	B.2	B.3	B.4	B.5	C.1	C.2
Call Success Rate (normal) %	35	53	57	39	59	59	45	43	34
Call Success Rate (attack) %	35	53	57	43	68	59	36	43	40

Table 6.4 Call success rates under attack 1 for 8, 17, 25, and 60 target calls.

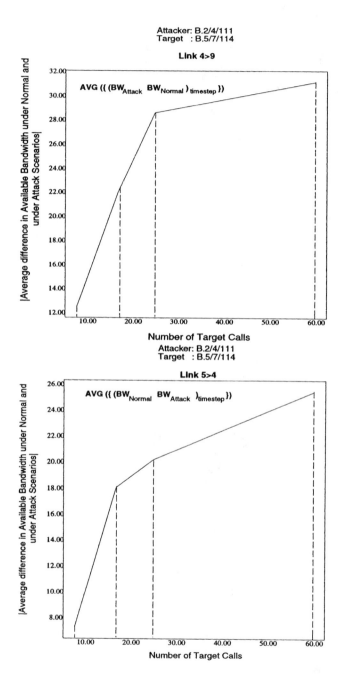

Figure 6.21 Difference in available bandwidth under normal and attack scenarios, averaged over all nodes, as a function of the number of target calls.

6.4.4 Attack 2

This attack constitutes a realization of basic attack 5, previously presented in subsection 6.3.3.5. It is restated as follows.

If a call is already established between any two nodes in the network, routed through an intermediate node already under the attacker's control, and if the source (or destination) node is on the perpetrator's list of "node(s) to be attacked," redirect all the traffic cells arriving on this channel to a periodically selected arbitrary channel that is already established. The attack applies to a call that may already be established or is currently under consideration. The attack is launched when and after the call has been successfully established.

Assume that the perpetrator controls node B.2 and its target is node B.3. The choice of B.2 as the location of the attacker follows the same reasoning as in the previous subsection 6.4.3. To maximize interception by the attacker, a node that is connected directly to B.2 is selected as the target node. While both B.3 and B.4 qualify, node B.3 is selected arbitrarily. Thus, calls originating at node B.3 and destined for C.1 or C.2 are more than likely to be routed through B.2 and may be subject to attack. These are labeled target calls. In this work, calls originating at the nodes of group C and destined for B.3 are also viewed as target calls. Also, the rate at which the outgoing VPI/VCI pair is modified for the target calls at B.2 is specified by the redirection period, in terms of timesteps. The complete list of channels at B.2's disposal includes B.2 → C.2, B.2 → B.3, and B.2 → B.4, which it will employ to deliberately misdirect traffic. Thus, the input file "attack.in" will contain the line: "2 111 112."

6.4.4.1 *Experiments.* For this attack, four cases are considered to develop an understanding of the impact of the two variables – number of target calls and frequency of redirection. For cases 1 and 2, the number of target calls is set at 30 and 50, respectively, and the redirection time period is selected at 5,000 timesteps. Cases 3 and 4 are designed to intensify the severity of the attack. The number of target calls is set at 80 and 120, and the redirection time period is decreased to 2,000 timesteps, implying faster hopping between the available channels for misdirecting the traffic cells. Input file "atm_input_5" is modified with the number of target calls, while input file "ATM_PARAMS" is modified to accept the values of the redirection period.

6.4.4.2 *Analysis of the Simulation Results.* To gain an understanding of the impact of this attack, key data from the simulation are logged, as explained subsequently.

First, when the perpetrator at the attacker node successfully intercepts a target call, the following data are written into the log file:

BEGIN
ATTACK 2 IN PROGRESS
target:100 attacker:101
[*ATT*2] init, count is 37
END
[106011] BW_AVAIL: link [2|3]:150631

[106011] ATT2 adding 32002:5 32004:4
new attack two added on vpci:32002
—ATT2— tagging [32004 : 4][32002 : 5]

The content of the "count" field represents the number of calls that have been processed by the switch up to the present time. The last line is important in that it implies that the call has been intercepted by the attacker. The original input/output vpi/vci pair for the cells of this call relative to the switch fabric consists of identifier 32004 at port 4 for the input side and 32002 at port 5 for the output side.

Second, upon selection of a randomly chosen arbitrary output vpi/vci pair, the following information is logged. It indicates that for the vpi/vci identifier 32008 at port 4 on the input side, the new output vpi/vci pair consists of identifier 1015 at port 4:

[214252] ATT2 NEW OUT: for input:32008 |4 orig:32006|5 old:32006|114230 new:1015| 4

Third, when a cell is dropped, the following data are logged, implying that at the time instant given by 480939 timesteps, the ATM cell along link 4 → 6 is dropped. The physical propagation delay along the link is 652, and the total bandwidth capacity is 156000, of which 3072 is available at the current timestep:

[480939] PACKET DROP output buffer link: LINK [4/6fd : 5st : 1] t/l 156000:652: bw:3072

The information contained in the logs is utilized for a comparative analysis of cell drop behavior in the four cases, under normal and attack scenarios.

Figures 6.22 and 6.23 present the cell drop behavior in each of the four cases, and focuses on the links that exhibit a difference between the normal and attack scenarios. For such links, the arrow identifies the link by direction, and the two sets of numbers represent the cells dropped under the normal scenario, indicated by the symbol N, and attack scenario, indicated by symbol A. It is pointed out that at any node, the output buffer is shared among all of the outgoing links.

Case 1:

For a total of 30 target calls and a redirection period of 5000 timesteps, differential cell drop behavior is observed at five links, B.2 → B.4, B.4 → B.1, C.2 → B.2, C.2 → C.1, and A.2 → A.1. Possible explanations are presented subsequently.

- Along the link, C.2 → C.1, cell drop is high under the normal scenario, implying that a number of calls are attempting to transport an excessive number of cells along this link. The significant reduction in cell drop under the attack scenario may be explained as follows. Given that the attacker is located at node B.2, for many of the calls originating at node B.3 and routed through B.2 toward the destination C.1, the cells are deliberately misdirected along other links. As a result, fewer cells are transported along the link C.2 → C.1 in turn implying less cell drop.

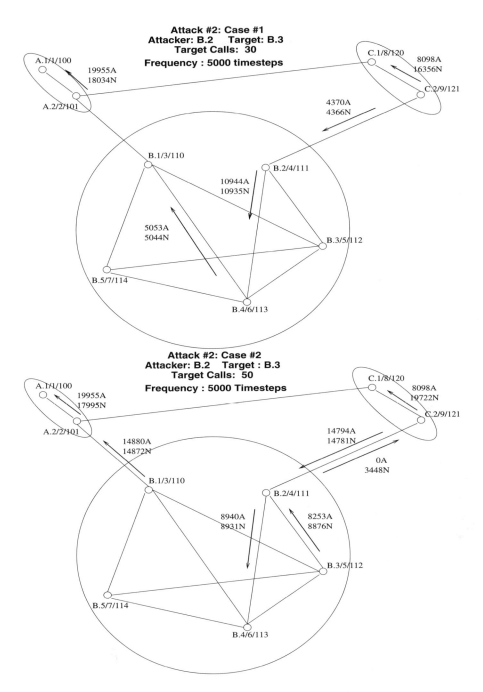

Figure 6.22 Cell drop rates at the links under normal and attack 2 scenarios.

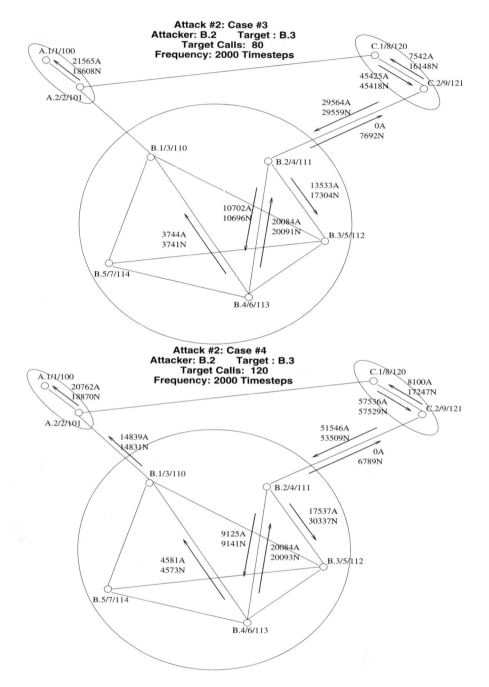

Figure 6.23 Cell drop rates at the links under normal and attack 2 scenarios (Cont'd).

One would expect link B.2 → C.2 to behave similarly to the link C.2 → C.1. However, there is no differential cell drop between the normal and attack scenarios along the link B.2 → C.2, the reason being as follows. Just as a subset of the cells meant for transport from B.3 to C.2 along link B.2 → C.2 are redirected elsewhere, similarly, a subset of the cells meant for transport from B.3 to B.4 along link B.2 → B.4 are possibly misdirected along B.2 → C.2, thereby canceling the two opposing effects. The calls originating at B.3, destined for B.4, and routed through B.2 may constitute a part of the traffic synthesized by the traffic generator within ATMSIM 1.0.

- Along the link A.2 → A.1, cell drop is high under the normal scenario. The fact that cell drop is high also along the link C.2 → C.1 possibly implies that a number of calls are established between nodes C.2 and A.1 and their constituent cells are being transported from node C.2 to node A.1. Under attack, the congestion along C.2 → C.1 is reduced, enabling more cells to be routed from C.2 toward A.1 via C.1. This contributes to more cell drop along the link A.2 → A.1 under the attack scenario. The slight increase in the number of cells dropped along the link B.4 → B.1 implies an increase in the number of cells being transported from B.4 possibly toward the nodes of group A. This may bear a slight contribution to the increased cell loss under attack along link A.2 → A.1.

- The difference in the cell drop along link C.2 → B.2 between the normal and attack scenarios, is slight. The overloading of the single buffer at node C.2 from excess traffic along the link C.2 → C.1 is more likely than the onset of the attack to influence the cell drop along the link C.2 → B.2.

- The less than significant increase in cell drop under the attack scenario along the links B.2 → B.4 and B.4 → B.1 may be attributed to the increase in the number of cells that are misdirected by B.2 toward B.4, some of which find their way toward B.1. Thus, the traffic along the links B.2 → B.4 and B.4 → B.1 increases, causing a slightly higher cell drop.

Case 2:

Relative to case 1, only the number of target calls is increased, to 50, in case 2. Except for the links B.2 → C.2 and B.3 → B.2, the general behavior of the links under case 2 is similar to that in case 1. The differential cell drop under normal and attack scenarios along links B.1 → A.2, C.2 → B.2, and B.2 → B.4 is slight, and as in case 1, they may be attributed to either the overloading of the buffers at the nodes or the misdirection of cells by the attacker node B.2. The behaviors along links C.2 → C.1 and A.2 → A.1 are similar to those in case 1, and similar explanations will apply. Under more target calls, the differential behaviors along these links are magnified, as revealed by the increased difference in the cell drop rates between the normal and attack scenarios.

Given the increased number of target calls in case 2, more of the cells from B.2 transported to C.2 via B.2 under the normal scenario are misdirected away from link B.2 → C.2 and possibly along link B.2 → B.4 under the attack scenario. The result is a significant differential cell drop behavior between the normal and attack scenarios along link B.2 → C.2, which, contrary to expectation, failed to manifest itself in case 1.

The appreciable decrease in the cell drop rate at link B.3 → B.2 under the attack scenario, may be explained as follows. Under normal scenario, nodes B.1, B.5, and B.4 transport a number of cells through B.3 towards B.2 or the nodes of group C. Also, a part of the traffic cells from C.2 or C.1 intended for nodes of group B may pass through node B.3. The resulting overloading of the buffer of node B.3 causes an appreciable cell drop along the link B.3 → B.2. Under attack, the increased number of target calls originating at B.3 implies that a significant number of cells originating at B.3 and intended for the nodes of group C via B.2 are misdirected at B.2 towards B.4. In turn, B.4 directs these misdirected cells towards B.5, B.1, and B.3. The resulting overloading of B.4's buffer implies a decrease in the number of cells sent by B.4 or via B.4 towards the nodes of group C or node B.2 via node B.3. Also, cells originating at the nodes of group C and intended for B.3 are intercepted and misdirected by B.2, thereby relieving pressure on the output buffer of B.3. Thus, the traffic encountered by B.3 decreases, in turn, implying less cell drop along the link B.3 → B.2.

Cases 3 and 4:

Under cases 3 and 4, while the number of target calls is increased to 80 and 120, the redirection time period is decreased to 2000 timesteps. The network continues to exhibit the same general behavior as in cases 1 and 2, the only difference being that the effects are amplified. In addition, the link B.2 → B.3 reveals an appreciable difference in the cell drop rate between the normal and attack scenarios, which may be explained as follows. Under the normal scenario, the increased number of cells stemming from an increased number of calls originating at the nodes of group C and destined either for B.3 or for other nodes of group B but routed through B.3 congest the link B.2 → B.3, causing an appreciable cell drop rate. Under attack, node B.2 redirects these cells away, thereby lowering the cell drop rate along the link B.2 → B.3.

In summary, the inference from cases 1 through 4 is that while the impact of attack 2 manifests itself in the network, the manifestation may occur close to the location of the perpetrator or at an unexpected region of the network, away from the location of the target node and the node controlled by the perpetrator. Thus, localizing the source of the attack may be difficult. In essence, attack 2 is highly dynamic, and its observable impact is a strong function of the current state of the network. Conceivably, attack 2 may be detected in an otherwise stable network by detecting a sharp change—increase or decrease—in the cell drop rate when the external traffic is intense. However, under light traffic load attack 2 may elude easy detection.

Under attack 2, cells intended for a node may appear at unexpected nodes, causing confusion and other unintended effects at higher levels. This study is limited in that the analysis is confined to the network layer and makes no attempt to systematically examine the receipt of stray cells and detect the attack through a comprehensive high-level analysis.

6.4.5 Attack 3

This attack constitutes an implementation of basic attack 6 discussed previously in section 6.3.3.6. It is restated as follows.

Effort is made to establish dummy sessions with other nodes in the current peer group and reserve network resources so as to deliberately deny service to genuine users. In developing this attack, the perpetrator may synthesize either a few bogus calls, each requiring significant bandwidth resource, labeled type A, or a number of bogus calls, each requesting modest bandwidth, labeled type B. Unlike attacks 1 and 2, this is a diffused attack in that it lacks a specific target node or link. Instead, it aims to inflict as much damage as possible to the entire network.

Assume that the attacker is located at node B.4. The choice is dictated by the fact that B.4 occupies a central place in the peer group with direct connection to all of its peer nodes in the current group. Although B.3 is also connected directly to all other nodes in the network, B.4 is selected arbitrarily over B.3. The reference input file "atm_input_6" is modified to include the dummy call SETUP requests along with the bandwidth resource request and their constituent traffic cells. In addition to its effort to establish bogus calls with its peer nodes B.1, B.2, B.3, and B.5, node B.4 will also attempt to establish bogus calls with nodes C.2 and A.2. Although not a gateway node and even if it is not a peer group leader, the PNNI 1.0 specifications permit node B.4 to possess knowledge of the gateway nodes of other peer groups.

6.4.5.1 *Experiments.* Logically, the routes of the bogus calls are chosen as follows: B.4 → B.1 → A.2 → A.1, B.4 → B.1, B.4 → B.2 → C.2 → C.1, B.4 → B.2, B.4 → B.3, and B.4 → B.5. A number of preliminary simulations were designed and executed to determine (i) the range of the bandwidth requests under types A and B and (ii) the number of bogus calls under each of types A and B to yield maximal differential network behavior under attack 3. From the results of the preliminary simulations, while bogus calls of type B request bandwidths in the range 4 Mb/s to 6 MB/s, calls of type A request bandwidths in the range 20 Mb/s to 30 Mb/s. A total of three sets of input traffic are synthesized, sets 1 through 3, and the number of calls in each set for each type of traffic is so chosen that the cumulative bandwidth requests for all bogus calls under each of types A and B are similar. Set 0 corresponds to the normal scenario and serves as a reference. Table 6.5 presents the details of the input traffic for each of the three sets. An increase in the number of bogus calls beyond set 3 reveals no significant change in the network behavior under attack.

6.4.5.2 *Analysis of the Simulation Results.* By nature, attack 3 aims at denying user calls. Thus, a logical measure of the success of this attack is the user call success rate. It may be pointed out that the graphs shown here present only the legitimate user calls, not bogus calls.

	Number of Bogus Calls	
Set #	Type A	Type B
0	0	0
1	12	60
2	24	120
3	60	300

Table 6.5 Details of the input traffic for attack 3.

Figure 6.24 plots the fraction of successful user calls in the entire network, expressed as a percentage of total user calls inserted into the network, as a function of the input traffic set, for each of type A and B attacks. For an increasing number of bogus calls from 0 (set 0) to 60 (set 3) for type A and from 0 (set 0) to 300 (set 3) for type B, the severity of the attack increases, causing a decrease in the call success rate. At the peak of the attack, virtually all of the available bandwidth is usurped by the bogus calls, and the network reaches a saturation point. Further increase in the number of bogus calls is accompanied by negligible impact on network performance. Comparative analysis of the call success rates for type A and B reveals that while the attack with type A bogus calls causes network performance to drop faster than with type B bogus calls, the call success rate in the entire network drops to its lowest level under type B bogus calls. The key reason is that

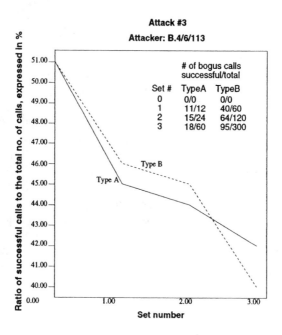

Figure 6.24 Fraction of successful user calls in the entire network, expressed as a percentage, as a function of the input traffic set, for type 3A and 3B attacks.

bogus calls with lower bandwidth demands are more likely to be established than calls with high bandwidth requirements, eventually usurping more of the network resources and inflicting greater damage on the network performance. In addition, the larger number of successful bogus calls under type B traffic implies that a greater fraction of the network's computational resources are consumed, thereby slowing the call setup process. It is also observed that despite its severity, attack 3 is unable to zero the call success rate in the entire network, reflecting the richness in route availability in representative networks and their resistance to complete defeat by a perpetrator.

To gain a better understanding of the nature of attack 3, Figures 6.25, 6.26, and 6.27 present the fractional call success rates at select individual nodes of the network, A.1, A.2, and B.5, respectively. The rationale underlying the choice of the nodes is as follows. Node A.1 is located farthest from the source of the attack. It is located in a different peer group and does not constitute a gateway node. While node A.2 is located in peer group A, different from group B where the perpetrator is located, it is connected to peer group B by virtue of being a gateway node. Node B.5 is situated in the same peer group as the attacker and is expected to bear the brunt of the attack. In Figure 6.25 the call success rate at node A.1 reveals no influence from the attack. This is expected, since the perpetrator at B.4 cannot reach node A.1. The gateway node, A.2, is visible to the attacker, and its susceptibility is reflected by the drop in the call success rate, shown in Figure

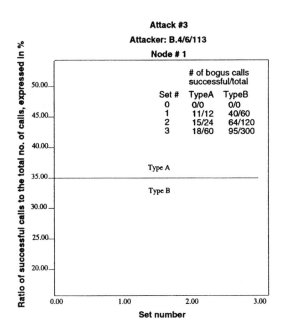

Figure 6.25 Fractional call success rate at node A.1 under attack 3, as a function of the input traffic sets.

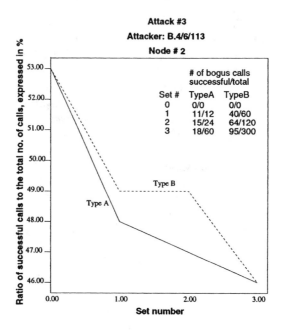

Figure 6.26 Fractional call success rate at node A.2 under attack 3, as a function of the input traffic sets.

6.26, from 53% to 46%, which equals 7%. In contrast, the drop in the call success rate at node B.5, shown in Figure 6.27, is 62% - 54% = 8% for attack with type A traffic and 62% - 52% = 10% for attack with type B traffic. Clearly, by virtue of being much closer to the attacker, node B.5 is more susceptible to attack 3 than A.2. To gain statistical confidence and for reliable simulation results, every experiment is repeatedly executed a number of times, and the final simulation result is obtained through averaging the data across all of the runs.

When a given network operating at the stable region experiences a sharp drop in the average call success rate throughout the network as well as at a number of the peer nodes of a group, and no obvious reason is present, it is likely to be a victim of attack 3. Where the bandwidth demands of a number of successfully established calls are appreciably high, the attack may have been launched with type A traffic. In the event that the call setup time is appreciably higher and the number of calls with high bandwidth demand is less than significant, the attack is likely to have been launched with type B traffic. Localizing attack 3 is relatively straightforward, since the nodes close to the source of the attack, especially in the same peer group, are affected more severely than those at a distance, particularly in other peer groups.

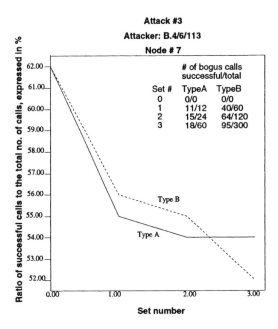

Figure 6.27 Fractional call success rate at node B.5 under attack 3, as a function of the input traffic sets.

6.4.6 Attack 4

This attack constitutes an implementation of the complex attack 8 discussed previously in section 6.3.3.8. Under this attack, a perpetrator controlling a node in the network will attempt to usurp all network resources at the disposal of a target node, thereby isolating it from the remainder of the network. The underlying mechanism employed by this attack is similar to that utilized in attack 3 (subsection 6.4.5) in that the perpetrator attempts to establish bogus calls to as many of the neighboring nodes of the target node as possible, routing them through the target node and usurping as much as possible the bandwidth resources of the links connected to the target node.

6.4.6.1 *Experiments.* To understand how the relative locations of the attacker and target nodes in the network topology influence the impact of this attack, especially in relationship to the connectivity with other peer groups, a number of attacker node and target node combinations are examined in this study. A total of three sets are included in the study here. In this subsection a target call is defined as one that includes the target node in its route, either as the source, destination, or any intermediate node. From the study described earlier in subsection 6.4.5, the use of type B traffic is observed to cause much greater damage to the network than type A and is therefore employed in this study for worst-case analysis. The appropriate traffic input files are modified to include bogus calls that utilize the routes specified for each of the three sets of attacker node and target node combinations.

Under set 1, B.3 and B.4 constitute the location of the perpetrator and the target node, respectively. The input file "attack.in" contains the line 1 112 113. For each of the four routes: B.3 → B.4 → B.1, B.3 → B.4 → B.5, B.3 → B.5 → B.4, and B.3 → B.4 → B.2 a total of 80 bogus calls are synthesized with bandwidth demands ranging from 4 Mb/s to 6 Mb/s. The rationale underlying the choice of the (B.3, B.4) node pair is as follows. Given that both B.3 and B.4 constitute the only nodes that connect to the gateway nodes in peer group B, many of the intergroup calls are likely to include them in their routes. The choice of any one of them as a target node offers a good chance to observe the attack in action. Here, B.4 is arbitrarily chosen as the target node, while B.3 is selected as the attacker node. It is pointed out that under the choices, the attacker node is directly connected to the target node.

Under set 2, while the attacker continues to be situated at B.3, the target is located at an interior node B.5 that is connected only to a single gateway node, B.1, and presumably encounters the least intergroup traffic of all other nodes in peer group B. As in set 1, the attacker node is directly connected to the target node. A secondary aim underlying this choice is to examine how well this attack refrains from disturbing the remainder of the network while focusing the damage on the relatively remote target node. The input file "attack.in" contains the line: 1 112 114. For each of the four routes, B.3 → B.5 → B.1, B.3 → B.5 → B.4, B.3 → B.4 → B.5, and B.3 → B.1 → B.5, a total of 80 bogus calls are synthesized with bandwidth demands ranging from 4 Mb/s to 6 Mb/s.

Under set 3, the goal is to study the impact of the attack on a gateway node including the extent of the damage inflicted to other peer groups. The target is located at B.2, and the attacker is placed at node B.4. While the perpetrator is connected to both gateway nodes, it is also directly connected to the target node. The input file "attack.in" contains the line: 1 113 111. For each of the three routes, B.4 → B.2 → C.2, B.4 → B.2 → B.3, B.4 → B.3 → B.2, and B.4 → B.3 → B.2 → C.2, a total of 80 bogus calls are synthesized with bandwidth demands ranging from 4 Mb/s to 6 Mb/s.

6.4.6.2 *Analysis of the Simulation Results.* Figure 6.28 presents the average call success rates at each of the nine nodes of the network, under normal and attack scenarios, corresponding to set 1. The rate is computed based on all user call requests that are inserted into the network. As expected, the call success rate at the target node 6 (B.4) is reduced significantly from over 62% to 22%. However, because the perpetrator must use the links at node 5 (B.3) to launch the attack, the call success rate for legitimate user calls at node 5 is also sharply reduced, from 62% to 30%. An important observation is that for nodes 7 (B.5) and 4 (B.2) the call success rates are lowered, implying that the links connected to B.5 and B.2 are exploited by the perpetrator to launch the attack. Nodes 8 (C.1), 9 (C.2), and 1 (A.1) are located far away from the source of the attack and reveal zero impact on their call success rates. The increase in the call success rates at nodes 2 (A.2) and 3 (B.1) is explained as follows. Given that many of the legitimate user

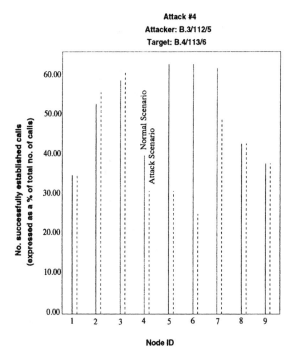

Figure 6.28 Call success rates at the nodes of the network for normal and attack 4 scenarios, for set 1.

calls fail due to lack of resources, the bandwidths along other links that would have been otherwise utilized under the normal scenario become available under the attack scenario. In turn, this permits a few of the calls that were previously denied under the normal scenario to be successfully established, causing a slight increase in the call success rates.

Figure 6.29 presents a detailed picture of the influence of attack 4 by focusing solely on the target calls, i.e., those that include the target node in their route. Calls with the target node as the source, calls sinking into the target node, and calls that simply pass through the target node all qualify. The graphs reveal that attack 4 is highly successful in that it affects every node in the entire network that attempts to reach the target node B.4. In contrast to Figure 6.28, target calls are solely considered here, and since they can fail only under attack, the graphs for all nodes under attack are uniformly lower than under the normal scenario.

The general behavior of the graphs in Figure 6.30 for set 2 is similar to that for set 1.

Figure 6.31 presents the success rates as a function of the nodes in the network for normal and attack scenarios, for set 3, for all calls in the network as well as the target calls. Except for the target node 4 (B.2) in Figure 6.31(a), the damage from the attack on other nodes in the network appears to be less severe than for

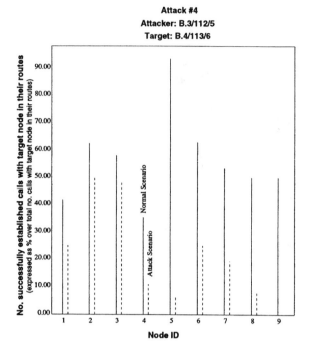

Figure 6.29 Success rates for target calls at the nodes of the network for normal and attack 4 scenarios, for set 1.

sets 1 and 2. Given its location in the peer group B and that it is a gateway node, B.2 appears to play a minor role relative to the intra-group calls within group B. As a result, an attack on it has not caused widespread reduction in the call success rates at all of the nodes of B. However, as shown in Figure 6.31(b), the target calls are adversely affected as a result of the attack. Clearly, inter-group calls from the nodes of group B to C are likely to be routed through B.2 and are susceptible to attack. Intergroup calls originating at the nodes of group C and destined for members of group B are not affected by the attack, since the link C.2 → B.2 is beyond B.4's capacity to attack. By definition of an ATM network and according to the PNNI 1.0 specification, a node within a peer group may not possess knowledge of the internal topology of a different peer group. This is shown in both Figures 6.31(a) and 6.31(b).

In summary, attack 4 can serve as an effective tool against a peer node, except when it serves as a gateway node. The target and attacking nodes are likely to experience the most significant drop in their call success rates, which in turn, may serve to point to the source of the attack.

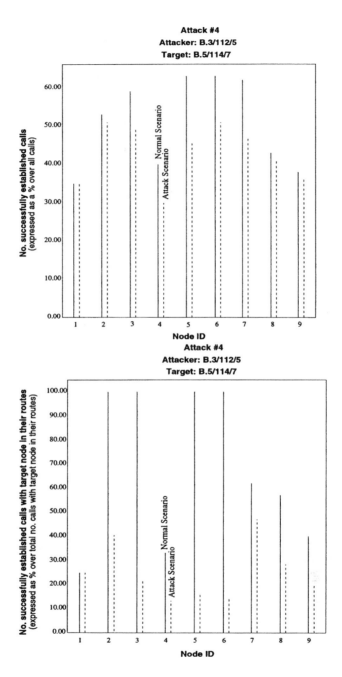

Figure 6.30 Success rates for (a) all calls and (b) target calls, at the node of the network for normal and attack scenarios, for set 2.

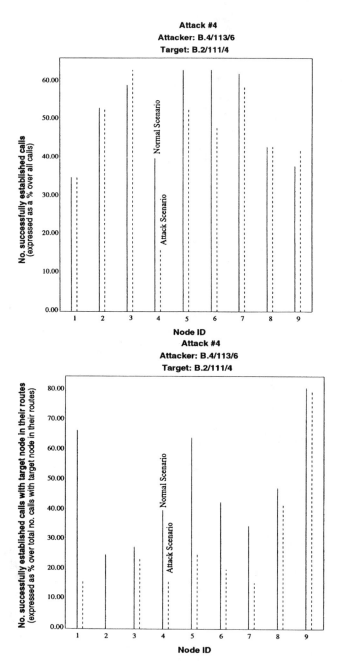

Figure 6.31 Success rates for (a) all calls and (b) target calls, at the node of the network for normal and attack scenarios, for set 3.

6.4.7 Attack 5

This attack constitutes a realization of the complex attack 9 discussed previously in section 6.3.3.9. The aim is to attempt to prevent two specific nodes from communicating with one another. As in the case of the previous attack 4 (subsection 6.4.6), here, type B traffic is employed. While the link interconnecting the two specific nodes is labeled target link, the calls routed through the target link are termed target calls.

6.4.7.1 *Experiments.* As with the attack in the previous subsection, to understand how the relative locations of the perpetrator and target links in the network topology influence the impact of this attack, especially in relationship to the connectivity with other peer groups, a number of attacker node and target link combinations are examined in this study. A total of three sets are included in the study here.

Under set 1, the perpetrator is assumed located at node B.4, and the target link is between B.3 and B.5. The corresponding link identifiers are 9 and 22. The routes for the bogus calls are chosen as: B.4 → B.5 → B.3 and B.4 → B.3 → B.5. For each of these two routes, a total of 100 bogus calls are synthesized with bandwidth demands ranging from 4 Mb/s to 6 Mb/s. Given its location, the link between B.3 and B.5 is likely to encounter few calls originating at other nodes and routed through it. The rationale for selecting the link is to examine how well this attack refrains from disturbing the remainder of the network while focusing its damage on the relatively remote target link.

Under set 2, the perpetrator is assumed located at node B.3, and the target link is between B.1 and B.4. The corresponding link identifiers are 4 and 17. The routes for the bogus calls are chosen as: B.3 → B.1 → B.4 and B.3 → B.4 → B.1. For each of these two routes, a total of 100 bogus calls are synthesized with bandwidth demands ranging from 4 Mb/s to 6 Mb/s. Observations indicate that a number of calls originating at B.1 and destined for the nodes of group C tend to utilize the link between B.1 and B.4. The goal of set 2 is to study the impact of attack 5 on relatively heavily utilized links.

Under set 3, the perpetrator is assumed located at node B.3, and the target link is between B.2 and C.2. The corresponding link identifiers are 11 and 24. The route for the bogus calls is chosen as: B.3 → B.2 → C.2, and a total of 100 bogus calls are synthesized with bandwidth demands ranging from 4 Mb/s to 6 Mb/s. The objective in this set is to study the impact of the attack on an inter-group link.

6.4.7.2 *Analysis of the Simulation Results.* The simulation results for each of the three sets are encapsulated through two graphs that plot the link utilization as a function of the link identifier. The link utilization for a link L is measured as the ratio of the number of successfully established calls (K_L) that are routed through L to the total number of calls inserted into the network (T_L) for which L is prescribed in their routes. Therefore, link utilization is given by $\frac{K_L}{T_L}$. In the ideal scenario, all T_L calls would be established successfully. Thus, link utilization reflects the behavior of the network relative to its ideal performance. While graph

1 takes into consideration all calls inserted into the network, graph 2 is confined to target calls.

Figure 6.32 presents the simulation results for set 1. As expected, in Figure 6.32(a), the link usage percentages for link identifiers 9 and 22 are significantly diminished under attack. However, the link usages for link identifiers 10 (link B.4 → B.5) and 21 (link B.4 → B.3) also reveal significant impact, reflecting their use by the attacker in synthesizing the bogus calls. The reason for the decrease in link usage for link identifier 20 (B.4 → B.2) is that a number of calls from B.5 to C.2 scheduled to utilize link identifier 20 fail under attack. Also, the failure of calls from B.5 to C.2 imply that more resources are likely to be available along the link B.2 → C.2. As a result, calls originating at B.3 and destined for group C through B.2 will incur a greater chance of success, thereby pushing the use of link identifier 19 (B.3 → B.2) higher. Figure 6.32(b) shows that only calls originating at B.5 and B.3 utilize link identifiers 9 and 22 and that they are severely affected by the attack.

Figure 6.33 presents the simulation results for set 2. Figure 6.33(a) reveals that the attack has adversely affected a few links beyond the target link identifiers 4 and 17. The wide sphere of influence of the attack is revealed more clearly in Figure 6.33(b), where for links 1, 2, 4, 7, 11, 14, 15, 17, 20, and 24 the link usage, based on target calls, is reduced significantly. This confirms the hypothesis that the target link between B.1 and B.4 is a popular route for calls originating at a number of nodes in the network.

Since the graphs in Figure 6.33(a) are based on all calls, not merely target calls, the influence of attack 5 is not clearly revealed. Under attack, many of the target calls fail, thereby leaving significant bandwidth along other links unoccupied. Other call SETUP requests that do not contain the target link in their route and may have failed earlier under the normal scenario now find available bandwidth resources and are successfully established. They, in turn, cause these link utilizations to climb higher, as seen in Figure 6.33(a). It is also observed that for links 5 (B.1 → B.5) and 26 (C.1 → A.2) the link utilization is higher under the attack scenario. The reason may be as follows. Consider calls originating at node B.1 and destined for node C.1 that utilize the target link B.1 → B.4 and fail under the attack scenario. This results in additional bandwidth available on link C.2 → C.1. Conceivably, calls originating at C.2 and destined for A.2 through C.1 that may have failed earlier under the normal scenario will now succeed, thereby causing an increase in the link utilization for link identifier 26.

Figure 6.34 presents the simulation results for set 3. As expected, in Figure 6.34(a), target link 11 (B.2 → C.2) reveals a sharp drop in link utilization. Despite being a target link, link 24 does not experience any reduction in link utilization, since the ATM network principles, as confirmed in PNNI 1.0, deny the perpetrator at node B.3 knowledge of the internal topology of peer group C. As a result, B.3 is precluded from synthesizing appropriate bogus calls. In addition, a significant drop in its utilization under attack implies that link 19 (B.3 → B.2) is the preferred route for calls from peer group B to C and is therefore susceptible to attack 5. Link 26 (C.1 → A.2) reveals a slight increase in its link utilization under attack,

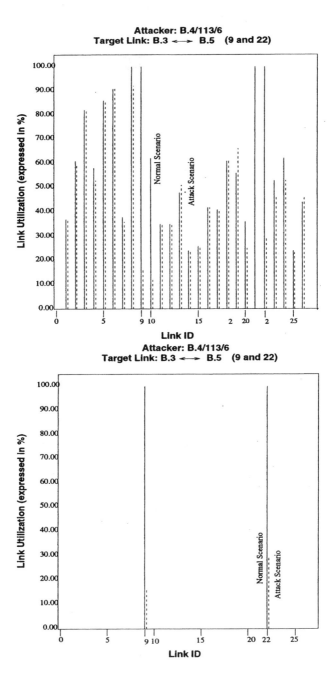

Figure 6.32 Link utilization as a function of link identifiers, for attack 5, set 1, for (a) all calls inserted into the network, (b) target calls.

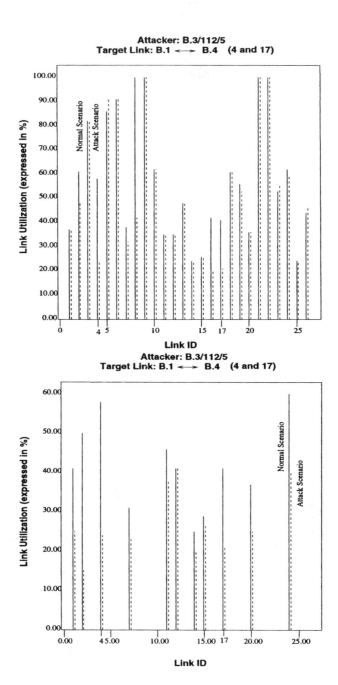

Figure 6.33 Link utilization as a function of link identifiers, for attack 5, set 2, for (a) all calls inserted into the network, (b) target calls.

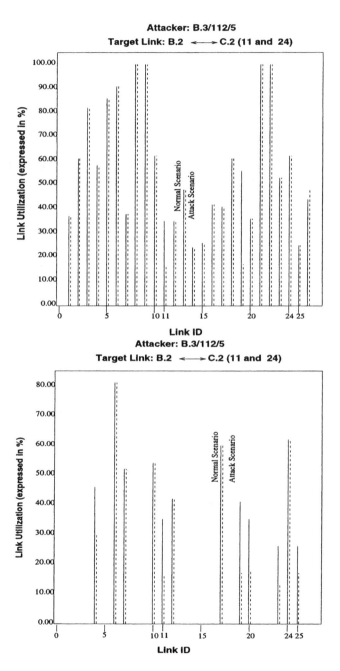

Figure 6.34 Link utilization as a function of link identifiers, for attack 5, set 3, for (a) all calls inserted into the network, (b) target calls.

which may be explained as follows. Under attack, calls originating in peer group B and destined for C.1 would have failed, thereby freeing bandwidth along the link C.2 → C.1. A few calls from C.2 destined for peer group A that may have been rejected earlier under the normal scenario due to inadequate bandwidth resources along the link C.2 → C.1 are now likely to succeed. In turn, this will cause link utilization along link 26 to rise.

Figure 6.34(b) reveals that in addition to the target link 11 (B.2 → C.2), links 4 (B.1 → B.4), 20 (B.4 → B.2), 23 (B.5 → B.4), and 25 (C.2 → C.1) experience a reduction in their link utilization values, measured by the number of target calls. With the bandwidth resource along B.2 → C.2 usurped by bogus calls under attack, user calls originating at B.1 and routed through B.4 and B.2 toward peer group C will fail, leaving behind unused bandwidth along links 4 and 20. Similarly, user calls originating at B.5 and routed through B.4 and B.2 toward peer group C will fail, leaving behind unused bandwidth along link 23. Furthermore, user calls originating at B.2 and routed through C.2 toward C.1, will fail, leaving behind unused bandwidth along link 25.

In summary, attack 5 can serve as an effective tool in inhibiting communication between two peer nodes, connected through a link. Where one of the nodes belongs to a different peer group, the effectiveness of the attack is greatly diminished. The target link and a few of the neighboring links that are employed in synthesizing and launching the attack are likely to experience the most significant drop in their link utilization rates, which may serve to point to the source of the attack.

6.5   Problems and Exercises

1. Identify significant vulnerabilities in IP networks and synthesize appropriate attacks.

2. Synthesize coordinated attacks, providing justifications for your design. Analyze the consequences of the attack and means of defeating them.

3. What key advantages do modeling and asynchronous distributed simulation provide in vulnerability analysis and attack modeling?

4. In ATM networks, although the payload portion of each ATM cell may be encrypted end-to-end and at the AAL layer, the header portion must be accessible to every intermediate ATM switch fabric so that the VPI/VCI pair may be modified. Explain whether this constitutes a vulnerability, and design counteracting measures, addressing the issue of network performance.

5. Obtain a copy of the documentary film *Hackers: Computer Outlaws*, aired by The Learning Channel (TLC), July 25, 2001, and find out the fundamental reasons that the Phone Phreaks were able to break into the Bell Telephone System.

7

Complex Vulnerabilities and Highly Sophisticated Attacks

In Chapter 6, vulnerability analysis focused on identifying the weaknesses of a network, starting with the basic principles that underscore the given network. In contrast, this chapter analyzes the key principles and assumptions that define the network itself, in an effort to discover the presence of fundamental weaknesses, if any, and the exact circumstances under which the network may be broken. Clearly, such vulnerabilities are highly complex and are likely to require sophisticated attacks. Consider IP networks, a key strength of which is resilience, stemming from the lack of a priori knowledge of the exact route of any IP packet. A key assumption underlying this resilience is that an IP router, upon intercepting an incident IP packet, forwards it in the general direction of its final destination node toward a subsequent IP node through the least congested link. When a perpetrator intercepts and deliberately discards an IP packet at an intermediate node, the assumption is abruptly broken. Given that the Internet has been in use for some time, the circumstances under which the fundamental assumptions break down are well known. However, for networks that are relatively recent, such circumstances may not be public knowledge, implying that someone, somewhere, may know precisely how to bring a network down.

The remainder of this chapter focuses on ATM networks and describes two recently discovered scenarios where network performance is severely degraded.

7.1 Influence of Source Traffic Distribution on Quality of Service

A principal attraction of ATM networks is that the key quality of service (QoS) parameters of every call, including end-to-end delay, jitter, and loss, are guaranteed by the network when appropriate cell-level traffic controls are imposed at the user network interface (UNI) on a per call basis, utilizing the peak cell rate (PCR) and the sustainable cell rate (SCR) values for the traffic sources. There are three practical difficulties with these guarantees. First, while PCR and SCR values are, in general, difficult to obtain for traffic sources, the typical user-provided parameter is a combination of the PCR, SCR, and maximum burstiness over the entire duration of the traffic. Second, the difficulty in accurately defining PCR arises from the requirement that the smallest time interval must be specified over which the PCR is computed, which, in the limit, will approach zero or the network's resolution of time. Third, the literature does not contain any reference to a scientific principle underlying these guarantees. Under these circumstances, the issue of providing QoS guarantees in the real world, through traffic controls applied on a per call basis, is rendered uncertain. This section aims at uncovering through systematic experimentation a relationship, if any, between the key high-level user traffic characteristics and the resulting QoS measures in a realistic operational environment. An experiment is designed that consists of a representative ATM network, traffic sources that are characterized through representative and realistic user-provided parameters, and a given set of input traffic volumes appropriate for a network-provider-approved link utilization measure. The key source traffic parameters include the number of sources that are incident on the network and the constituent links at any given time, the bandwidth requirement of the sources, and their nature. For each call, the constituent cells are generated stochastically, utilizing the typical user-provided parameter as an estimate of the bandwidth requirement. Extensive simulations reveal that for a given link utilization level held uniform throughout the network, while the QoS metrics, end-to-end cell delay, jitter, and loss are superior in the presence of many calls each with low bandwidth requirement, they are significantly worse when the network carries fewer calls of very high bandwidths. This finding underscores a vulnerability of ATM networks that a perpetrator may exploit to launch a performance attack.

The experimentation and analysis described in this section, leading to the discovery of the vulnerability in essence constitutes the sophisticated attack.

7.1.1 Introduction

One of the principal attractions of ATM is that despite being a high-speed network, the key quality of service (QoS) parameters of every call, including end-to-end delay, jitter, and loss, are guaranteed by the network when appropriate cell-level traffic controls are imposed on a per call basis. In ATM, such traffic controls may not be located in the switches deep in the network. The reasons are as follows. First, they would slow down the normally fast flow of traffic through the switch fabric appreciably. Second, ATM cells from different traffic sources are statistically multiplexed in the switches deep within the network, and isolating them by user calls in order to selectively apply controls on those that violate the traffic contract may be very difficult at least. Third, it is unreasonable, from a computational perspective, to expect the ATM switches deep within the network to be aware of the details of the QoS contract between the user and the network.

Therefore, traffic controls are necessarily located at the user network interface (UNI), where each individual user negotiates traffic contracts with the network. While the computational burden associated with monitoring traffic on a per call basis is greatly reduced, high bandwidth traffic sources pose a significant challenge to implementing traffic controls [106].

7.1.1.1 *QoS Guarantee in ATM Networks: Underlying Assumptions.* Delay constitutes a primary QoS metric. For a given ATM cell, the delay incurred during its transit through the network is a function of many parameters, the most important being traffic volume, buffer architecture, and buffer management. Figure 7.1 depicts a representative probability distribution of cells incident at a buffer of a ATM node, corresponding to a given, i^{th} connection. While the delay incurred by the cells range from D_{min} to D_{max}, the maximum delay that the underlying application may tolerate is generally lower that D_{max} and is represented by the symbol, D^*_{max}. Thus, cells that fall beyond D^*_{max} are viewed as too late and considered lost. Additional cells of the connection may also suffer loss from buffer overflow. The area under the graph in Figure 7.1, between the boundaries D^*_{max} and D_{max}, reflects the fraction of the total number of cells lost due to excessive delay and is expressed by 10^{-x}, where x is defined by the application. Thus, if an application can tolerate 0.001% cell loss stemming from excessive delays and buffer overflow, i.e., one cell lost in 100,000, then x is at most equal to 5. The fraction of the cells successfully transported up to the ATM node in question would be given by $1 - 10^{-x}$.

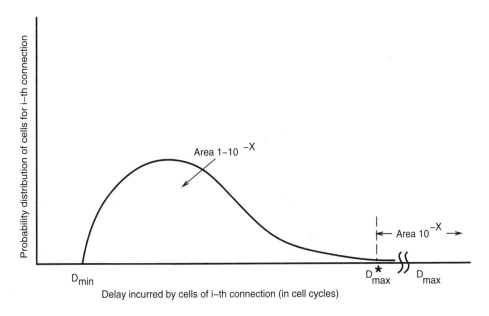

Figure 7.1 Probability distribution of cells at a buffer of an ATM node.

A key assumption in ATM is that the principal QoS parameters of each call, including end-to-end delay, jitter, and cell loss, are guaranteed by the network when proper off-line dimensions are designed for the switch buffers and policing buckets and appropriate cell-level traffic controls are imposed on a per call basis, utilizing the peak cell rate (PCR) and sustainable cell rate (SCR) values of the traffic sources. The controls generally involve the call setup process, scheduling of the cells at the switches, cell-level admission control, priority control, policing, and traffic shaping. The ATM Forum's technical committee [107] formally states the assumption on page 35 of the "Traffic Management Specification, Version 4.0" as follows: "By controlling the connection traffic flows, allocating adequate resources and selecting suitable routes, a network may provide the requested service category and QoS parameters for CLP = 0 and CLP = 1 cell flows."

Analysis reveals that this assumption faces three practical difficulties. First, while true PCR and SCR values are, in general, difficult to obtain for traffic sources [108], the typical user-provided parameter is a combination of the PCR, SCR, and the maximum burstiness of the traffic over its entire duration. A reliable SCR value requires measurement over a long period of time, which is difficult unless details of the traffic's characteristics are known a priori. Second, the difficulty in accurately defining PCR arises from the requirement that the smallest time interval must be specified over which the PCR is computed, which, in the limit, will approach zero or the network's resolution of time. This issue is elaborated subsequently. Third, the literature does not contain any reference to a scientific principle underlying this guarantee.

The definition of PCR is given by the following. If N is the number of cells generated by a traffic source or received at a switch over a time interval T, then

$$\mathrm{PCR} \;=\; \lim_{T \to 0} \frac{N}{T}$$

In reality, the smallest value of T is given by the resolution of time of the entity defining or computing the PCR. Conceivably, a given user's, say A's, notion of the resolution of time may be different from that of another user, say B, and still different from that of the switch that constitutes the network. In general, for a traffic source, the rate at which traffic is generated influences its notion of the resolution of time. Thus, the problem is that the PCR value computed by A may be very different than if it is computed by B, with the consequence that there is an unintended impact on the switches, as explained subsequently. Recall that the idea behind the PCR is that the UNI will impose controls on the incoming cells of a user, utilizing the negotiated PCR value as a threshold. If more cells arrive at any time than are permitted by the PCR value, the traffic control will detect and rectify the situation by either dropping user cells immediately or marking them as candidates to be dropped in the future by the network in the event of congestion.

As an example, suppose that user A sends a maximum burst of 100 cells spread over 1 ms, shown in Figure 7.2(a). Assume that the resolution of time for the user is 1 ms, i.e., the smallest time window it can comprehend is 1 ms. This yields a PCR value of $\frac{100 \times 53 \times 8 \text{ bits}}{1 \text{ ms}} = 42.4$ Mb/s. From the user's point of view, the PCR value of 42.4 Mb/s is meaningful and logical. Imagine that the traffic controls at the UNI

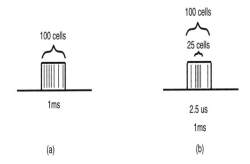

Figure 7.2 Practical difficulty with PCR-based traffic controls.

monitor A's traffic and find no violation. Assume further that in the cell burst that originated at A and propagated to the switch, a subburst of 25 cells occupy a time window of 2.5 μs, as shown in Figure 7.2(b). In general, the resolution of time of a switch is much finer than that of the individual traffic sources, which implies that the switch perceives the PCR of the incoming traffic as $\frac{25 \times 53 \times 8 \text{ bits}}{2.5 \times 10^{-6} \text{ s}} = 4.24$ Gb/s. Thus the switch's perception of the incoming traffic's PCR is significantly higher than that permitted by the contract negotiated between the source traffic and the network. While this may cause excessive competition for buffers and buffer management computational resources, leading to poor QoS metrics, the problem is fundamentally due to the different resolutions of time.

7.1.2 Review of the Literature on QoS Guarantees and Traffic Characterization

7.1.2.1 *QoS Guarantees Through Traffic Controls on a Per-Call Basis.* In ATM networks, the key reason for negotiating a contract with a traffic source is to offer it QoS guarantees through traffic controls exercised on all traffic sources in the network [107]. According to the literature, network providers have exploited a number of different traffic controls [109] including resource reservation for use during the transport of traffic, appropriate scheduling strategies at multiplexers and switches, and proactive congestion control techniques such as call level admission control, burst level admission control, peak-rate and sustainable-rate policing, and traffic shaping. Other competing techniques include reactive congestion control methods such as window-based or source-rate controls, back-pressure-based on/off flow controls at the UNI, assigning cell loss priorities, deliberate blocking of bursts, and explicit congestion notifications. Of these, the leaky bucket algorithm constitutes the simplest and most popular traffic control mechanism. However, its performance has been shown to decrease [110] drastically when it attempts to police bursty sources such as video codecs. The window-based controls include jumping window, triggered jumping window, and exponentially weighted moving average (EWMA) mechanisms. For a comprehensive list of policing mechanisms, the reader is referred to [111][112]. A fuzzy-logic-based policer has also been used in traffic control [113]. In essence, virtually all of the schemes providing QoS guarantees through traffic controls are based on some form of queue management.

Kurose [114] discusses open issues and challenges in providing QoS in high-speed networks. In the literature, two schemes involving communication abstractions have been proposed. Ferrari's "real-time channels" [115] and Zhang's "flows" [116] seek to provide QoS guarantees through dynamic scheduling on a per packet basis. The central idea in [115] is that an individual call may be accepted when upon evaluation of the characteristics of the resulting aggregate traffic it is found that the delay requirements of the call in question will be satisfied. The algorithm bases its decision on the knowledge of the current traffic characteristics, network capacity, maximum service time in the node for the packets, call traffic parameters, and QoS parameters. It presupposes that the user is capable of identifying and supplying the parameters of his connection even before the network has decided on whether or not to admit the call. Furthermore, the decision to admit or not occurs at each node, independent of all other nodes. This poses a problem in that the traffic characteristics deep within the network may be very different from that at the network boundary, due to statistical multiplexing. Zhang [116] proposes a virtual clock traffic control scheduling algorithm for packet networks, where resources are preapportioned. The algorithm provides a method to control the average transmission rates of the sources and helps bound the delay.

In reactive QoS management schemes, policing is carried out by requiring some sources to transmit information at low data rates. Resources are not allocated to individual connections. Proactive QoS management schemes achieve their goals by managing resources at call setup time. A source declares its QoS needs, and the network reserves resources, where possible. If resources are unavailable, the network rejects the call, invoking admission control policies. These schemes utilize priority scheduling or other mechanisms to schedule cells at the multiplexers and switches, and rate control policing.

The Tenet group's approach [117] to real-time communication is connection-oriented and involves the reservation of resources during the connection establishment phase. In the forward pass, the best possible resources are reserved. Later, in the reverse pass, the demand for the best resources is relaxed, taking into account the client's QoS requirements, the actual QoS achievable under the current network conditions, and reservation requests in the forward pass. To satisfy packet delay and jitter constraints for real-time communication in wide area networks and internetworks, the approach utilizes a performance-oriented call admission algorithm. An approximate traffic model with X_{min} (minimum packet inter arrival time), X_{ave} (average packet interarrival time), and S_{max} (maximum size of the packet) is used. A multiclass earliest due date (EDD) scheduling mechanism is employed in the network nodes for packet transmission. A distributed rate control algorithm that increases the deadlines for packets that arrive too soon serves as the flow control algorithm. Clark [118] adopts a predictive approach wherein the traffic characteristics of a source are predicted at call admission time and assumed to hold true for the entire duration of the call. Clark presents a weighted fair queuing scheduling algorithm to guarantee QoS on packet networks. Lazar and Pacifici [119] present a resource control algorithm that is based on asynchronous time sharing (ATS) and in which cooperative scheduling is based on traffic prediction. It is based on a fixed menu of QoS. Each real-time connection obtains from the network performance guarantees corresponding to those of its class. In a real-time scheduling algorithm (MARS), Hyman et al. [120] propose the concept of "schedulable region" of a queuing system, which refers to an intersection of the space of loads from different sources for which the QoS can be guaranteed. They also propose a real-time algorithm that attempts to increase the width of the schedulable region. Peha [121] proposes an integrated scheduling and admission control algorithm termed "priority token bank." Kreoner et al. [122] present an analytical analysis of the end-to-end delay and conclude that control mechanisms must focus only on the cell loss probability. Golestani [123] proposes a "stop and go queuing" approach in which packets arriving at a node during any frame are prevented from becoming eligible for transmission until the end of the frame. Thus, traffic characteristics are preserved, although extra delays are introduced into the system because of this wanton stoppage. Gong and Parulkar [124] suggest using an explicit rate control coupled with window control to limit the flow of a source to guarantee its QoS. They argue that rate control ensures compliance of sources relative to the requested bandwidth, while window control provides end-to-end speed matching. They claim that this will help the network perform effective management of resources and thus provide QoS guarantees. Aras et al. [125] present a survey of network architectures and protocols that support real-time services in packet-switched networks. They also discuss traffic characteristics and performance requirements for real-time applications. Lutas et al. [126] describe an architecture to interconnect an Ethernet to an FDDI network via an ATM backbone using satellite communication. They report that the challenges encountered in such an interconnection include segmentation and reassembly of LAN frames and buffer management in the network. A key limitation of all of these efforts is that they have not been validated through actual implementation and, as a result, they fail to yield a practically realizable approach to guaranteeing QoS in networks.

Aurrecoechea et al. [127] provide a general survey of end-to-end QoS architectures for distributed multimedia systems. Boudec et al. [128] propose a flow control mechanism for available bit-rate traffic. This approach utilizes measurements of the current queue length, bandwidth availability, and activity of the traffic sources to impose limits on the transmission rate of traffic sources. While Chang and Thomas [129] present a survey of different mechanisms to compute the effective bandwidth for resource allocation and QoS guarantees in ATM networks, Choudhury et al. [130] argue that effective bandwidth computing algorithms fail to capture the effects of statistical multiplexing. Tryfonas et al. [131] propose new estimation algorithms to compute effective bandwidth for stored MPEG video streams. Lakshman et al. [132] present feedback-based rate-adjustment algorithms for the transport of video sources. Resynchronization problems stemming from cell delay variations and solution mechanisms are presented in [133][134].

7.1.2.2 *Traffic Characterization*. Characterizing traffic is important, for it helps us to understand network performance through analytic modeling as well as through behavior modeling and simulation. The central issue is how to model the arrival process of cells generated from bursty sources to either a multiplexer or switch. The issue remains an open question despite a growing literature in ATM traffic modeling. In general, ATM is expected to carry continuous bit rate (CBR) traffic, variable bit rate (VBR) traffic, and unspecified bit rate (UBR) traffic. CBR is defined by a constant rate of flow of digital information over a prolonged period, an example being a 64 kb pulse code modulated (PCM) digital voice. VBR is characterized by a peak bit rate data flow, with a long-term average bit rate value that is less than that of the peak. Real-world examples include compressed video transfer and bursty data flow.

Frost and Melamed [135] suggest describing a source traffic through three components, A_n, B_n, and W_n, where A_n refers to the interarrival time, i.e., the length of the time interval separating the arrival of the nth cell from that of its predecessor, B_n is termed "batch size" and is the number of cells all of which arrive at the same time instant T_n, and W_n is the amount of effort that the system must expend to process the nth arriving cell. For example, W_n is the frame size, in bits, of a video source. According to ITU-T (previously CCITT) [136], three parameters are necessary to characterize a source, p, the peak cell rate or the peak arrival rate of the cells of a source; m, the average cell arrival rate, averaged over a long interval; and β, the burstiness or the ratio between the peak cell rate and the average cell rate. CCITT [136] provides typical parameter values for mean burst lengths and average cell arrival rates for representative audio, video, and data sources. In the literature, the models frequently chosen for analytical modeling and traffic generation include the two-state on–off model, Poisson process, Bernoulli processes, Markov models, Markov modulated deterministic or Poisson processes where the state of the Markov chain determines the rate at which the correlated cells arrive, transition modulated processes [136][135][137], the binomial and autoregressive model [138], autoregressive moving average model [139], discrete-time Markovian arrival process model, linear model for bit rate prediction [140], geometrically modulated deterministic model [141], and fluid-flow model [142]. Leland et al. [143] propose a self-similar model to characterize highly superposed LAN traffic. Plotkin and Roche [144] label the degree of randomness in cell scattering entropy and use it as a traffic descriptor. To capture the burstiness of a traffic source such as compressed video, Knightly and Zhang [145] introduce a parameterized traffic model, D-BIND, in which traffic is specified through multiple worst-case rate-interval pairs over a given time interval. The literature reports efforts to superimpose different traffic sources on a link to reduce the dimensionality problem.

Video sources are expected to dominate ATM networks in terms of the number of sources and the net traffic volume. Examples include cable TV, video telephones, and video conferencing [146]. The quality of the traffic generated by these sources is dependent on the nature of the picture, relationships between the frames, required QoS, and the encoding technique employed.

To transport voice sources over ATM networks [147] a number of different techniques are employed. These include PCM coded 64 kb/s CBR transmission, transmitting speech in a compressed mode using digital speech interpolation and cell discarding, G-764 packet voice protocol, continuous generation of cells for voice, and superposition of multiple voice sources.

Traditionally, data sources including terminal emulation and file transfers have been represented through the Bernoulli model, train model, or on–off model. In the Bernoulli model, a cell is generated at the beginning of each slot with probability p. Where the interarrival times between cells are assumed exponential, the model degenerates to a Poisson process. When cells are generated in batches, the model assumes the form of a compound Poisson process. In the train model [148], cells are assumed generated in trains during the active periods. The on–off model may be viewed as an interrupted Bernoulli model with a succession of active and silent periods, each of which may be geometrically distributed.

All of the reported studies incur the following limitations. The first is the absence of comprehensive experimental measurements to help validate the proposed traffic models for different sources under different applications. Second, all of the models assume, without scientific justification, that the source is able to characterize its traffic into peak cell rate (PCR), sustainable cell rate (SCR), and maximum burst length, which in reality is very difficult. Third, virtually all of the models address the traffic at the lower, cell level. The literature on source-level traffic distribution is sparse.

7.1.3 The Influence of Number and Nature of Sources and the Source Traffic Bandwidth Distribution on QoS for a Uniform Link Utilization

7.1.3.1 The Underlying Thinking and Rationale. Network performance [149][150] may be viewed either from the perspective of each user or from that of the network provider. A user is interested solely in the QoS of his/her own traffic. In contrast, the network provider cares about two factors. The first is to maximize the link utilization in the network, since links constitute a significant investment, and the second is to ensure the QoS guarantees for every user's traffic, thereby maintaining overall customer satisfaction. From the literature, it is evident that traffic controls at the cell level cannot be relied upon to deliver QoS guarantees. In this book, it is reasoned that the next logical step consists in applying traffic control at the call level, focusing on a high-level source traffic parameter. A reasonable choice is the source traffic's bandwidth requirement. Thus, the investigation takes the form of assessing the impact of source traffic bandwidth distribution on the QoS. Next, a hypothesis is submitted that the number of traffic sources and the bandwidth distribution of the source traffic, admitted subject to a specific network criterion, to be determined subsequently, will exert a unique influence on the cumulative delay distribution at the buffers of the representative nodes and hence on the QoS guarantees of each call. As shown through Figure 7.1 earlier, QoS guarantees are inversely proportional to cell delay distributions. The underlying thinking is as follows. The cumulative buffer delay distribution, at any given node and at any time instant, will clearly reflect the cumulative effect of the traffic distributions of the multiple connections that are currently active on the input link. Furthermore, the bounds of the cumulative buffer delay distribution at the nodes of the network clearly dominate the QoS bounds of each of the constituent source of user traffic. Thus, for each individual traffic source, the buffer delay distributions at the nodes of the network, obtained for different traffic distributions, will serve as its QoS measure. Returning to the issue of source traffic bandwidth distribution, it is evident that the distribution can assume innumerable forms. This gives rise to the need of calibrating it, preferably to tie it to an important network characteristic. The average "link utilization" over the entire network reflects an overall measure of the utilization of the network, and since it is constitutes an important system parameter from the network provider's point of view, the investigation is constrained by a set of reasonable choices of "link utilization" factors.

During the study, the average "link utilization" computed over all the links in a network is held at a choice value through high-level call admission control, i.e., by limiting the volume of the incident traffic on the network at any time. The study then focuses on determining how the two QoS metrics, cell delay and cell delay variation (jitter), as well as the loads at the buffers of a node, behave for different combinations of number of traffic sources and source traffic bandwidth distributions. The QoS parameter cell loss is a manifestation of delay beyond the maximum acceptable delay point of any connection on the delay distribution curve. Thus, the nature of the impact on cell delay will extend into cell loss.

7.1.4 Experimentation

The independent variables for the experiments include (a) a representative network topology, (b) user traffic characteristic expressed in the form of a bandwidth requirement that reflects a combination of the PCR, SCR, and maximum burstiness over the entire duration of the traffic, (c) frame duration, (d) deadline within which the message should reach the destination, (e) the destination host name, and (f) details of call arrival distributions, namely, the phase, duration, active interval, silence interval, and cell arrival times.

Figure 7.3 presents the first of two representative network topologies used in this study. The study started with a network consisting of 2 switches, and the experiments were repeated with three, four, five, and six switches. For all of these cases, the general nature of the behavior of the results remains unchanged. With increasing network size, the complexity of developing the simulator, debugging, and the length of the wall clock time for each simulation run increases significantly. In the interest of completing the work in a reasonable time frame, without any loss in either accuracy or generality, the representative network topology consists of six switches. To obtain representative and meaningful results, each simulation is first permitted to run for a length of time to complete initialization and eliminate transients in the network. Then, the simulation is executed for at least one million ATM traffic cells, requiring upwards of approximately 8 hours of wall clock time. The determination of each traffic point requires at least 5–8 trial simulation runs.

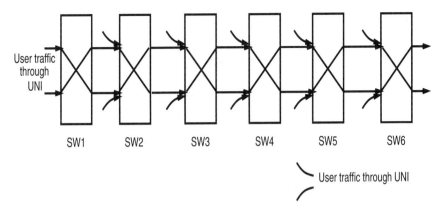

Figure 7.3 Representative network topology 1.

A number of experiments were conducted where each experiment captures the behavior of a QoS metric for a series of traffic points. A traffic point represents a combination of a specific number of traffic sources, their bandwidths, and the nature and parameters of the underlying cell-level traffic at the different nodes of the representative network, obtained through trial experiments, such that the resulting average link utilization, i.e., over all the links in the network, is held at a predetermined value. Through a number of experiments, described subsequently in subsection 7.1.6.3 of this chapter, it was observed that the general nature of the results is similar for different choices of the average link utilization value. Thus, without any loss in generality, the average link utilization value was set at 0.8 in this study, which, though arbitrary, reflects a reasonable choice. While the parameters for cell-level traffic generation are obtained from the literature, as described in subsection 7.1.2.2, Table 7.1.4 describes the traffic points, i.e., the input source traffic distributions, that are used in the experiments, all of which yield an average link utilization of 0.8. In Table 7.1.4, column 2 represents the traffic point with the fewest number of traffic sources, 25, and virtually all are very high bandwidth sources. In contrast, column 9 represents the traffic point with the greatest number of traffic sources, 2040, all very low bandwidth sources. The traffic points between columns 2 and 9 reflect increasing numbers of traffic sources with decreasing bandwidth requirement. The underlying cell-level traffic for all of the traffic points does not incur any traffic control violation at the corresponding UNIs.

In the simulation the traffic generation process models traffic sources where the cell arrival pattern varies from constant bit rate to highly bursty. A bursty source may either transmit cells up to the peak rate for a variable amount of time, followed by a period during which no cells are transmitted, or generate cells on a continuous basis with the rate varying with time and occasionally reaching the specified peak rate. Three types of traffic sources are employed: variable bit rate

Type of Source	25 Sources		47 Sources		70 Sources		196 Sources		560 Sources		1120 Sources		1680 Sources		2040 Sources	
	No.	BW Mb/s	No.	BW Mb/s	No.	BW Mb/s	No.	BW Mb/s	No.	BW Mb/s	No.	BW Mb/s	No.	BW Mb/s	No.	BW Mb/s
Video1	1	40.0	1	13.0	5	11.0	14	4.929	40	1.60	80	0.743	120	0.48	160	0.595
Video++1	1	55.0	1	65.0	5	13.0	14	4.929	40	1.60	80	0.743	120	0.48	160	0.595
Video2	2	55.0	4	15.0	5	13.0	14	4.929	40	1.60	80	0.743	120	0.48	160	0.595
Voice1	2	0,064	4	0.064	5	0.064	14	0.064	40	0.064	80	0.064	120	0.064	160	0.064
Voice2	2	0.064	4	0.064	5	0.064	14	0.064	40	0.064	80	0.064	120	0.064	160	0.064
Data1	2	30.0	4	30.0	5	1.2	14	0.5	40	0.17	80	0.1	120	0.08	160	0.05
Data2	2	30.0	4	30.0	5	1.2	14	0.5	40	0.17	80	0.1	120	0.08	160	0.05
Data3	2	30.0	4	30.0	5	1.2	14	0.5	40	0.17	80	0.1	120	0.08	160	0.05
Data4	2	30.0	4	30.0	5	1.2	14	0.5	40	0.17	80	0.1	120	0.17	160	0.05
Data5	2	30.0	4	30.0	5	1.2	14	0.5	40	0.17	80	0.1	120	0.08	160	0.05
Voice3	2	0.064	4	0.064	5	0.064	14	0.064	40	0.064	80	0.064	120	0.064	160	0.064
Video3	2	70.0	4	18.0	5	11.0	14	4.929	40	1.6	80	0.743	120	0.48	160	0.095
Video4	2	70.0	4	18.0	5	13.0	14	4.929	40	1.6	80	0.743	120	0.48	160	0.095
Video++2	1	60.0	1	65.0	5	13.0	14	4.929	40	1.6	80	0.743	120	0.48	160	0.095

Table 7.1 Source traffic points corresponding to an average link utilization of 0.8.

sources with random frame duration, variable bit rate sources with fixed frame duration, and constant bit rate sources with fixed interarrival gap between the cells throughout the duration of the calls. A total of six stochastic variables, conforming to either uniform, exponential, deterministic, or a variation of null distributions, as explained subsequently, are employed in generating traffic for a source. The variables include (1) call arrival time that represents the arrival time of a traffic source during the simulation period, (2) connection phase that encapsulates the phase at which a source starts to transmit following the connection setup, (3) call duration that represents the duration of the connection during the simulation, (4) active interval time, which reflects the time duration of a cell burst within a frame, (5) silence interval time, that implies the time period within a frame during which no cells are transmitted, and (6) cell arrival time, which encapsulates the arrival pattern of cells within the active period.

Audio traffic sources are modeled as on/off sources with random frame duration, with a peak bit rate of 64 kb/s, and choices of active and silence interval durations to yield a mean rate of 25.175 kb/s. While the active interval follows a distribution given by "10 ms + exponential (1.187, 0.0)," the silence interval follows a distribution expressed as "200 ms + exponential (1.646, 0.0)."

Video sources are modeled as variable bit rate (VBR) sources with 33 fixed frames per second, with the H.261 algorithm utilized for video compression. The choice of the silence interval distribution is such that it yields a mean rate of 1.28 Mb/s and a peak bit rate (PBR) of 3 Mb/s. The frames being fixed, the silence interval is nil.

In modeling high-definition television (HDTV) sources, the connection arrival distribution of the source is derived from a deterministic distribution function by modifying the latter's period ("per"). The modified function is utilized to generate a number of batches of video connections, with each batch consisting of a variable number of connections. In this study the connections of the first batch are characterized by a period of 0.0. For the second batch, while the first connection is characterized by a period incremented by "per," the remainder of the connections in the batch are labeled with period 0.0. As a result, all of the connections of the second batch arrive at the same time. Thus, unlike video connections that are characterized by arrival distributions, when connections with period 0.0 are encountered, the originating traffic source is immediately recognized as HDTV type.

Data sources are bursty, and the active and silence intervals are chosen so as to yield a mean rate of 1 Mb/s and PBR of 2 Mb/s. While the active and silence interval durations, both conforming to uniform distributions, are defined by the range {100 ms, 1.0 s} per frame, the sources are assumed to have an arrival phase that follows a uniform distribution in the range {0.0 s, 2.0 s}.

The dependent variables in the study constitute the performance metrics and consist of the following measures: (i) minimum, maximum, mean, and standard deviations of cell delay, (ii) jitter and absolute jitter in all of the queues—real-time video queue, real-time audio queue, and non real-time data queue, in every node of the network. In addition, the study collects and reports statistics on the (iii) minimum, maximum, mean, and standard deviation of cell loads in each of the queues in every node, and (iv) cell distribution function for each cell delay cycle.

7.1.5 Modeling the ATM Network, Simulator, and the Simulation Testbed (ATMSIM)

The two representative ATM networks in Figures 7.3 and 7.11, shown subsequently, are modeled and simulated in ATMSIM that utilizes as its core simulation engine the uniprocessor, event-driven simulator CSIM18 [151]. CSIM18 is a library of generic C++ classes and procedures implementing the underlying simulation operations. Elements of the ATM network are modeled through C++ modules that use CSIM objects and functions to produce estimates of time and performance of the dynamic behavior of the network. An ATM cell is encapsulated as a process that requests service at facilities that include links, multiplexers, and buffers. The processing of an ATM cell by a facility represents an event. The simulation clock maintains the simulation time. It starts at 0 and advances monotonically with the progress of the simulation, in terms of the cell cycle time. A "cell cycle time" serves as the basic unit of time, and its value is computed as 2.831 μs for an ATM layer link speed of 149.760 Mb/s corresponding to a physical layer SONET/SDH (OC-3) speed of 155.52 Mb/s. The performance statistics reported here are obtained in terms of cell cycle time units, while the connection intervals are measured in seconds.

In modeling the ATM network, the architecture of ATM switches, shown in Figure 7.4 is assumed to consist of input buffers. A set of three buffers serve the real-time video traffic, real-time audio traffic, and non-real-time traffic.

In Figure 7.4, as cells arrive asynchronously from the traffic sources, a cell, arriving at any given time slot is assigned to the corresponding buffer, provided that there is available space. The server utilizes the scheduling algorithm to first examine, then accept, a cell from the three buffers, and finally transmit an appropriate cell to the output link. It operates in discrete time steps, equal to the transition time of a single 53-byte ATM cell, 2.827 μs for a given 155 Mb/s link.

The simulations were executed on a stand-alone Sun Sparc 20 workstation, and each simulation run required 1.5 hours of wall clock time, corresponding to an actual network operation time of 1 s. A total of 300 simulation runs were executed for debugging and initial trials corresponding to different choices of the nature of the traffic distributions and different link utilization factors. The performance data collection and verification and reverification of the simulation results, required a cumulative total of 3800 hours of wall clock time.

7.1.6 Simulation Results and Analysis

The simulation data underlying all of the results reported here were obtained from the buffers of switch 4, deep inside the network. Experimental observations, reported later in subsection 7.1.6.2, indicate that the nature of the data at other switches in the network is similar to that at switch 4.

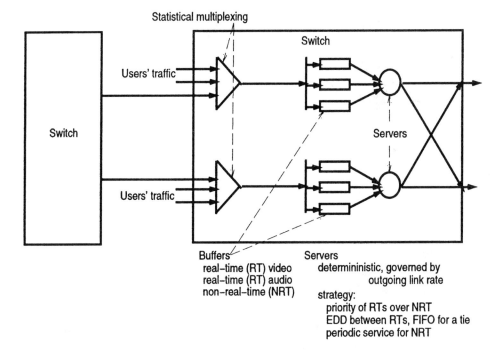

Figure 7.4 Modeling the switch in ATMSIM.

7.1.6.1 *Impact of Source Bandwidth Distribution on Cell Delay.* In the experiments the "cell delay" is the measure of the time difference between the instant the cell is generated, represented by the "time stamp," and the clock time when the same cell is released from the switch into the output link. The minimum and maximum delays refer to the smallest and greatest delays, respectively, acquired by any of the cells while being transported through a given link during the period of simulation. The average of the delays incurred by all of the cells transported through a given link in the simulation time period is the mean delay. Table 7.2 presents the maximum, mean, and standard deviations of the delays incurred by the cells carrying video traffic, corresponding to each of the traffic points as shown in Table 7.1. Figure 7.5 presents a graphical plot of the maximum delay, in cell cycles, incurred by the cells corresponding to real-time video connections, with the x-axis representing the traffic points of Table 7.1. A cell cycle corresponds to the length of time occupied by a 53-byte ATM packet in a 155.5 Mb/s ATM link, and measures at 2.872 μs. Figure 7.5 reveals that minimum and maximum delay values are at 5500 and 1000 cell cycles, respectively, and occur for 25 and 2040 traffic sources. Furthermore, the cell delay drops sharply as the number of sources utilizing the link increases. The abrupt drop in the cell delay occurs when the number of sources is 70. Thereafter, as the number of sources increases, up to 2040, the cell delay decreases gradually.

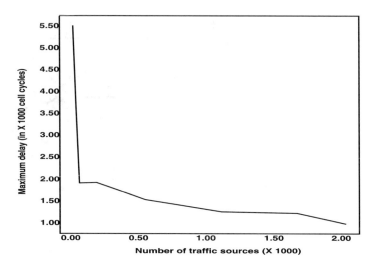

Figure 7.5 Maximum delay behavior of cells carrying video traffic, observed at the buffers of switch 4 for average link utilization of 0.8.

To verify whether the behavior reflected by the maximum delay variation is representative, Figure 7.6 plots the mean delay of cells carrying video traffic as a function of the traffic points. The nature of the behavior is observed to be similar to that for the maximum delay, including the fact that the graph drops sharply to 300 cell cycles at the traffic point consisting of 70 traffic sources. The highest and lowest mean cell delay values of 1200 and 220 cell cycles occur for 25 and 2040 traffic sources.

Clearly, the cell delay behaviors in Figures 7.5 and 7.6 represent the net behavior for all of the traffic sources transported through switch 4, and they serve as a bound for each of the constituent traffic sources. To fully understand the implications of

No. of calls	max delay (cell cycles)	mean delay (cell cycles)	std dev (cell cycles)
25	5555	1221	1123
47	3797	905	817
70	1963	317	338
196	1973	571	467
560	1584	567	451
1120	1303	478	357
1680	1280	458	348
2040	1035	252	212

Table 7.2 Maximum, mean, and standard deviations of the delays incurred by the cells carrying video traffic, observed at the buffers of switch 4 for average link utilization of 0.8.

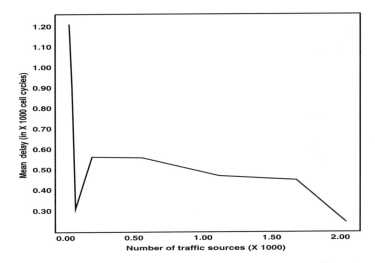

Figure 7.6 Mean delay behavior of cells carrying video traffic, observed at the buffers of switch 4 for average link utilization of 0.8.

the behavior in Figures 7.5 and 7.6, consider a histogram of cell distribution versus cell delay, in cell cycles, for different traffic points, shown in Figure 7.7. The x-axis represents the delay in cell cycles incurred by the cells in the real-time video buffer at switch 4, while the y-axis represents the number of cells incurring specific delays. To emphasize the most significant elements of the histogram, the x-axis is limited to delays of 15 cell cycles, although its range extends to 1035 cell cycles for the case of 2040 traffic sources to 5555 cell cycles for 25 traffic sources. Also, the results in Figure 7.7 are shown for three representative traffic points: 25, 70, and 2040 traffic sources. It is clear from Figure 7.7 that a much greater volume of cells is included within the 15-cell cycle delay for the traffic points with 70 and 2040 sources as opposed to that for the traffic point with 25 sources. Clearly, each of the cell delay distribution graphs, corresponding to a traffic point represents the net behavior of all of the traffic sources transported through switch 4 and serves as a bound for each of the constituent traffic sources. QoS guarantees are inversely proportional to the cell delay behavior. The higher the cell delay, the worse the QoS guarantee. Therefore, to achieve a superior cell delay QoS metric, given the desired link utilization factor of 0.8, the traffic point with 70 sources represents an excellent combination of number of traffic sources and source traffic bandwidth distribution. The findings may be easily extended to the cell loss QoS metric, given that cell loss is a manifestation of cell delay, beyond the maximum acceptable delay point of any connection on the delay distribution curve.

To serve as additional validation of the general nature of the results in Figures 7.5 and 7.6, Figure 7.8 presents the standard deviation (SD), in cell cycles, of the cell delay as a function of the traffic points. The standard deviation value is observed to be approximately one-fourth that of the mean delay. The SD for the traffic point with 25 sources is a high of 1123 cell cycles, and it drops sharply to

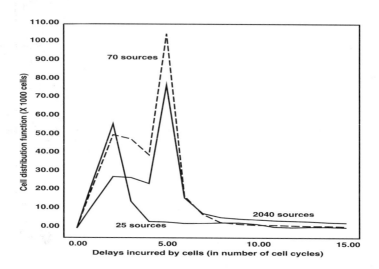

Figure 7.7 Histogram of cell distribution versus cell delay, expressed in cell cycles and limited to 15 cell cycles to emphasize the significant component of the distribution.

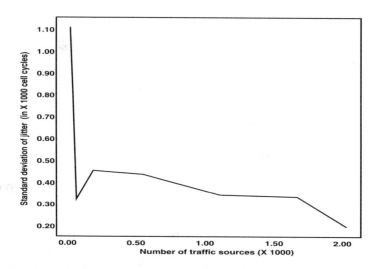

Figure 7.8 Behavior of the standard deviation of cell delay for cells carrying video traffic, observed at the buffers of switch 4 for average link utilization of 0.8.

338 cell cycles for the traffic point with 70 sources. However, the behavior of the standard deviation is similar to that of the maximum and mean cell delays. The low SD value for the traffic point with 70 sources also implies superior QoS in that although some cells may incur a large cell delay, approaching or even equal to the maximum delay, they are very few in number.

Analysis of the results indicates that when an ATM network carries a few very high bandwidth traffic sources, traffic cells incur excessive delays, implying poor QoS guarantees. The QoS improves dramatically as the traffic distribution moves to a higher number of traffic sources of lower bandwidth.

7.1.6.2 *Consistency of Cell Delay Behavior Across Different Switches in the Network.* To confirm that the behavior of the cell delay, cell loss, jitter QoS metrics, and buffer occupancies as observed at switch 4 is representative of the entire network, Figure 7.9 presents a comparative plot of the cell delay graphs incurred by the video sources at switches 4, 5, and 6. While the actual delay values differ, the nature of the behavior is similar across the different switches.

7.1.6.3 *Impact of Source Bandwidth Distribution on the Delay of Cells Carrying Video Traffic, Under Different Link Utilization Choices.* Figure 7.10 describes the influence of the different link utilization choices on the nature of the impact of source bandwidth distribution on the cell delay QoS metric. The choice of link utilization factors, from 0.6 to 0.96, constitutes a reasonable range over which network providers would prefer to operate their networks. The graphs corresponding to the five link utilization factors utilized are similar, corroborating the findings that the number of traffic sources and the source traffic bandwidth distribution bear a significant impact on the cell delay, jitter, and cell loss QoS metrics, and

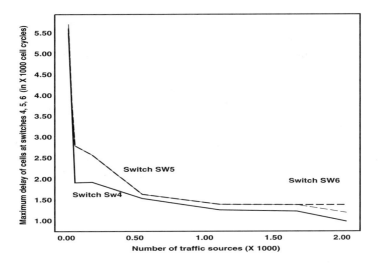

Figure 7.9 Maximum delay incurred by cells carrying video traffic, for switches 4, 5, and 6, observed at the real-time video buffers for average link utilization of 0.8.

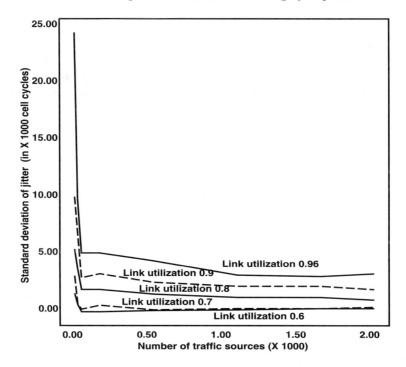

Figure 7.10 Maximum delay of video cells for different link utilizations.

that fewer high-bandwidth sources influence the QoS metrics adversely. As the network is pushed for higher efficiency, reflected by higher link utilization values, the QoS suffers. From Figure 7.10 it also follows that the adverse effect on the QoS metrics diminishes for choices of lower link utilization values. This inference is logical, since a low link utilization value implies less congestion from fewer traffic sources with low overall bandwidths.

7.1.6.4 *Impact of Source Bandwidth Distribution on the Delay of Cells Carrying Video Traffic, Under Different Active Interval Distributions.* To assess whether different values of active interval distributions will influence the findings, all of the previous experiments were repeated for different choices of values for the active interval. To achieve this objective, a new set of traffic points was obtained, in a manner similar to that described earlier in section 7.1.4, holding the link utilization at 0.8. Analysis of the data from these experiments, not shown here, reveals similar findings with respect to cell delay, cell loss, and jitter QoS metrics and buffer occupancies.

7.1.6.5 *Impact of Source Bandwidth Distribution on Network QoS for a Second Representative Network Topology.* While the switches in the representative network topology I are connected in tandem, the representative network topology II, shown in Figure 7.11, permits a choice of paths for source and destination users connected to switches SW1 and SW6. All of the experiments are repeated for the

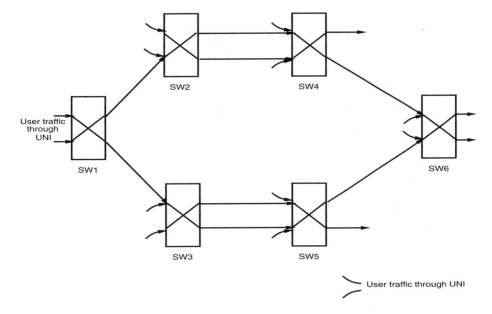

Figure 7.11 Representative network topology 2.

new topology, and the results relative to cell delay, cell loss, and jitter QoS metrics as well as buffer occupancy are similar to those obtained for topology I.

7.2 Susceptibility of the Traditional ATM Output Buffer Architecture to UNI-Compliant Traffic

A unique characteristic of ATM networks is the quality of service guarantee, which assures that for every user the negotiated QoS parameters, including cell throughput, cell delay, cell loss, and jitter, will be honored throughout the duration of service. To achieve this goal, ATM imposes two requirements. First, in the call setup process, for a call to be accepted, adequate resources, especially bandwidth, must be available and allocated at every link throughout the route, from the source to the destination. Second, during subsequent cell transport, the traffic must be UNI-compliant; i.e., it must conform to the negotiated parameters as monitored at the user network interface (UNI). Research findings, presented here, reveal that under certain traffic conditions and system parameter choices, QoS guarantees may be broken, and severe cell loss may occur, even when all ATM requirements are met. In essence, this research may be viewed as a highly sophisticated attack design. The principal sources of the problem include the traditional output buffer architecture in ATM switches, bursty traffic, and the lack of proven principles to estimate network traffic as well as to derive buffer sizes from traffic estimates.

The traditional output buffer architecture in ATM switches is realized in one of two forms, shared or separate. In the shared design, all of the N output links utilize a large common buffer, of size S cells, in an effort to achieve efficient use of the total buffer space. Under the separate buffer architecture, distinct buffers, each of size $\frac{S}{N}$ cells, are assigned to every output link, the goal being to realize fair buffer usage by every individual output link. For both architectures, however, the buffer organization is determined, permanently, at the time of the switch fabric design, and may not be altered during actual operation. The highly dynamic and stochastic nature of ATM traffic poses a key difficulty for this tradition. First, the distribution of the destination nodes of users' calls are unknown a priori, i.e., before the start of the network operation. As a result, for any given ATM switch, during operation, one or more of its output links may periodically incur stress from relatively very high traffic load. Second, given the bursty nature of the traffic, at the output links under stress the net traffic may exceed the link capacity for momentary periods of time, causing cell drop stemming from buffer overflow. Although ATM mandates very low cell loss rates, at under 0.001%, fast-growing link capacities beyond 155.5 Mb/s and approaching 9.1 Gb/s imply billions of cells transported per minute across a representative ATM network, and even the low 0.001% cell loss rate will imply a significant number of dropped cells, in the hundreds of thousands. Increasing the total buffer size may not constitute a viable solution due to the limited silicon area in the IC chips and also because buffer access times will increase significantly.

This section will focus primarily on the failure of the QoS guarantees, especially cell loss, as a consequence of the traditional buffer architecture. However, to facilitate the study, traffic cells are synthesized utilizing state-of-the-art traffic models, the volume of high-level call SETUP requests is set based on a logical principle that is explained subsequently, and the buffer sizes are copied from commercial ATM switches.

7.2.1 Introduction

Given the statistical nature of the traffic, buffering in any packet switch is unavoidable [152]. The underlying reason is that two or more packets or cells within the same time slot may be simultaneously destined for the same output port, even if the switch is characterized as output nonblocking. Under these circumstances, to avoid cell drop from buffer overflow, buffers must be associated with every output port of the switch. In addition, when the traffic arrival is bursty, i.e., abrupt changes in the number of cell arrivals occur for very short durations, the issue of cell drop is further aggravated, and this requires even larger buffers to achieve the same cell drop probability, compared to a scenario of uniform traffic arrival. Although the introduction of buffers of infinite capacity promises to eliminate all cell drop and imply superior performance for any traffic type, in reality, buffer sizes are finite. The traditional output buffer architecture comes in two forms, separate buffer and shared buffer. Under the separate buffer architecture, distinct and dedicated buffers are assigned to each output link of any switch. The result is fairness in buffer availability among the output links, yielding efficient behavior in the presence of bursty traffic. However, under intense traffic, when one or more buffers of specific output links are full, cells are dropped despite available space in the buffers of other output links. In contrast, the use of a shared buffer architecture, where all output links share a common buffer, significantly reduces the amount of memory required to achieve a given cell loss performance under uniform traffic [153][154][155][156][157][158], regardless of the buffer management strategy employed. Pashan, Soneru, and Martin [158] observe that for an 8X8 switch with output buffering and a net buffer of capacity 8,000 cells, under input traffic with an average burst length limited to 5 cells, and cell loss probability limited to 10^{-1}, while the shared buffer architecture sustains up to 88% of the load, the separate buffer design can carry only up to 45% of the load. However, Causey and Kim [159] report that when the ratio of the average burst length to the number of buffer spaces per output link increases, the shared buffer is often filled up to 80–90% with cells corresponding to a single output link, resulting in severe congestion and gross unfairness. In conclusion, while the shared and separate buffer architectures appear to function well for nonbursty and relatively bursty traffic, respectively, neither one of them is an ideal match for bursty input traffic.

7.2.2 Modeling the Traditional Output Buffer Architecture

The traditional output buffer architecture is modeled for a representative, 15-node ATM network, which in turn is derived from the very high performance backbone network service (vBNS) topology. The vBNS, shown in Figure 7.12, has been sponsored by the National Science Foundation and implemented by MCI [160]. The 15-node network is shown in Figure 7.13, with Figure 7.13(a) representing the high-level hierarchy with major US cities constituting the peer group leader nodes and Figure 7.13(b) presenting the constituent nodes of the individual peer groups. Each peer group consists of three nodes resembling a major city flanked by minor cities within a 40 mile radius, and the nodes are connected through intra-group links. The shaded node represents the designated peer group leader.

The underlying ATM node consists of N outgoing links, each with its associated separate buffer. Given that each of the N incoming and N outgoing links are rated at a speed of V b/s in this section, the core of the switch must sustain a cell transfer rate given by

$$T_t = \frac{cell\ size}{N \times V} = \frac{2.73\mu s}{N},\tag{7.1}$$

where the cell size is 53 bytes and the link speed V is 155.5 Mb/s. For an 8 X 8 switch fabric, the time to propagate from an incoming link to an outgoing link is 342 ns.

To ensure a representative and practical size for the separate output buffers, the state-of-the-art ForeRunner ASX-200 and ASX-1000 [161] ATM switches are analyzed to reveal an output buffer size of 13,312 ATM cells. This implies a total output buffer capacity of ($N \times$ 13,212) cells, given that the total number of buffers equals the number of outgoing links, N.

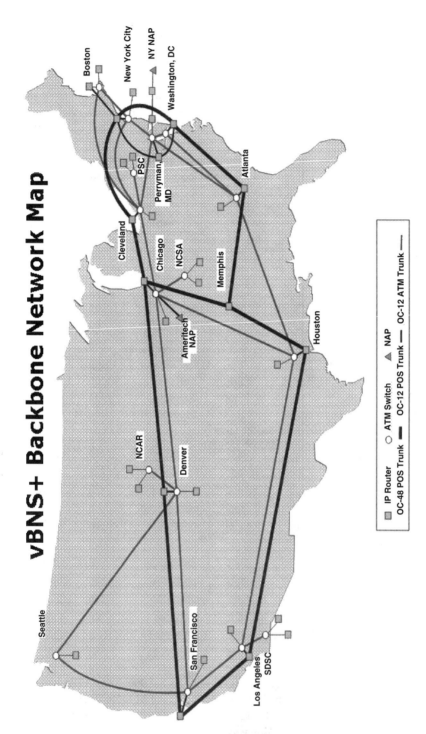

Figure 7.12 vBNS backbone ATM network topology.

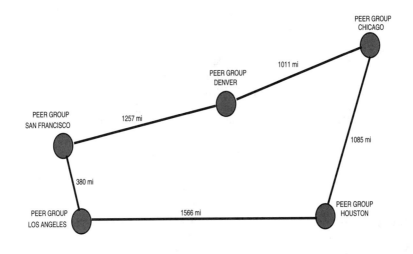

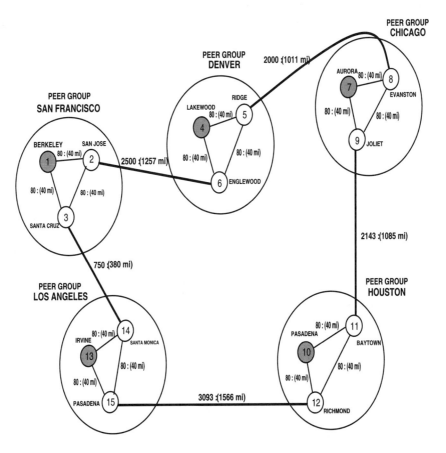

Figure 7.13 Representative 15-node ATM network.

7.2.3 Implementing the Traditional Output Buffer Architecture for a Representative ATM Network

The key details of the traditional output buffer architecture including the hardware and timing parameters were modeled and incorporated into an accurate distributed ATM network simulator that was executed on a network of Linux workstations, configured as a loosely coupled parallel processor testbed. The use of the asynchronous distributed simulation technique coupled with the concurrent processors executing asynchronously closely resembles an operational ATM network and is expected to yield realistic results. The resulting simulator consists of 20,000+ lines of C/C++ code and was executed for the 15-node representative ATM network shown in Figure 7.13. A large number of simulation experiments were designed and executed, each requiring approximately 40 hours of wall clock time. The cumulative size of the simulation data generated across the distributed processors, for each of a number of scenarios described subsequently in section 7.2.5, exceeds 2 Mb, and they were processed to yield unique insights.

The resolution of time for the simulator was set at 2.73 μs, reflecting the size of an ATM cell of 53 bytes and the fastest link speed of 155.5 Mb/s. This is referred to as the timestep. The accuracy of the simulator is defined by the choice of the 2.73 μs timestep, the correct order of execution of the events in the simulator, and the fidelity of the behavior models of the ATM switches and call processing nodes. Input traffic is inserted into the network in the form of stochastically generated high-level user calls at every node of the ATM network and the constituent traffic cells, audio, video, and data, corresponding to each of the user calls. The issue of traffic generation is addressed subsequently in subsection 7.2.4.

7.2.4 Input Traffic Modeling and Synthesis of Input Traffic Distributions

The choice of the input traffic and other network input parameters in network simulation is crucial for obtaining realistic performance data and useful insights into network behavior. The call processing architecture is distributed and employs 10 processors. Traffic generation represents a careful tradeoff between the goal of exposing the network to worst-case stress and examining its behavior and the need to ensure a stable operational network, one that executes continuously, 24 hours a day, 365 days a year, without failing. While much work has been reported in ATM traffic modeling, the literature on call setup requests is sparse. Given the lack of large-scale operational ATM networks in the public domain, operating under the mode of switched virtual circuits, actual data on call setup requests from operational networks are difficult to obtain. In this research, while the synthetic traffic is stochastic, it is designed to resemble an operational system. The key input traffic parameters include (1) call arrival distribution, (2) bandwidth distribution in the calls, (3) call duration distributions in the calls, (4) traffic mix, i.e., the relative percentage of inter- and intra-group calls in the entire network, and (5) low-level traffic models. With the exception of item 4, the distributions in items 1 through 5 were generated stochastically, and the choice of the key parameters is explained subsequently. Following their generation, traffic stimuli were saved in files for use during simulation.

7.2.4.1 *Call Cluster Arrival Distribution and Network Stability Criterion.* To emulate a network under intense stress, calls are assumed to arrive in clusters, at times defined by the interarrival interval. The number of calls in the clusters was stochastically generated, subject to a maximum of 25. To determine an appropriate call arrival distribution for the representative network modeled, call clusters were generated through a Poisson distribution function, and the network was simulated for different choices of the mean value of the distribution. For each simulation experiment, graphs of the call setup times as a function of simulation time, for every pair of source and destination nodes, were obtained and analyzed. Where any of the graphs exhibit a nonuniform behavior with the call setup time increasing consistently with the progress of simulation, the network was considered to be driven into an unstable region by the excessive call arrival density. The argument is that for networks to remain operationally stable, a graph of the call setup time as a function of simulation time, for any given pair of source and destination nodes, must remain uniform. Where all of the graphs exhibit uniform behavior, the network was considered to be within the stable operating region. Through trial and error, i.e., by executing a number of simulations for different choices of the mean of the Poisson distribution, this study yielded a call cluster arrival distribution that stressed the network to the edge of stability.

In this study user calls constitute one of two forms of traffic, data and voice/audio, key traffic parameters of which are presented in Table 7.3. While the parameters for audio traffic were obtained from [162], the choice of the parameters for the data traffic is explained as follows.

We argue that the trend of short call durations, of ftp, http, and email message type, will gain increasing prominence in the future, fueled by our impatience and the rapid growth in the link bandwidths that is already outpacing the expected increase in the average message size transported across the networks. Trace analysis shows that the current Internet traffic is already dominated by ftp and http transfers of data and images and that their dominance is increasing. We reason that while the ubiquitous T-1 line of the recent past, rated at 1.5 Mb/s, constitutes the lower bound of the user bandwidth requirement, the 25–80 Mb/s bandwidth rating of high definition TV (HDTV) constitutes the upper bound. Between these two extremes, the assumption of 20 Mb/s bandwidth availability for the average user call in the future is logical and reasonable. Assuming a typical 1 Mbit file size for ftp [95], the 20 Mb/s bandwidth translates into a 50 ms call duration. In this study the representative call duration for ftp was assumed to be 50 ms. How-

Service	Transmission Rate (bps)	Call Duration (seconds)
Voice/Audio	64K	90
Data/File Transfer	1M-8M	0.054

Table 7.3 Parameters of the traffic constituting user calls.

ever, to achieve higher call density in the network, i.e., to accommodate more user calls while maintaining backbone link bandwidth at 155.5 Mb/s, the user bandwidth distribution was set at 1–8 Mb/s, yielding an average of 4 Mb/s. While calls originated at all nodes in the network with uniform distribution, the destinations of the calls were stochastically generated. Last, 90% of all call requests involved data traffic, while 10% related to audio, and user calls followed a distribution of 80% intragroup and 20% intergroup calls. Stability analysis, utilizing the above parameters, reveals an intercluster arrival rate of 18 ms for the stability criterion point, which was utilized in the remainder of the experiments. Figure 7.14 plots the distribution of call setup times for different sets of {source, destination} node pairs, {1, 2}, {8,9}, and {10, 12}, as a function of the progress of simulation, corresponding to the stability criterion point. The graphs in Figure 7.14 are representative of the call setup behavior between other source and destination node pairs in the network.

7.2.4.2 Cell Level Traffic Parameters. The lower-level traffic models utilized in this study may be explained as follows. A two-state Markov process describes the audio traffic model with ON/OFF periods of activity that utilize the basic parameters provided in [163]. The mean of the active period interval (ON) is 352 ms, while that of the silent period interval (OFF) is 650 ms. The intercell duration within the ON period is 2112 μs, and the maximum burst length is 150 cells. The ftp-type data traffic generator utilizes an ON/OFF Markov chain model with the basic parameters extracted from [162]. The maximum burst length is 2,000 cells.

7.2.5 Simulation Experiments, Results, and Performance Analysis

The key objective of this section is to assess the performance of the traditional ATM switch fabric design and thus a simulation corresponding to the traditional ATM node that is developed. The synthetic input traffic is based on the assumptions of lower session durations and higher average user bandwidths, and both of these trends are likely to be widely prevalent in the future.

The output metrics in this study include the (1) throughput, i.e., the percentage of user cells successfully transported through each node of the representative ATM network, (2) average throughput over all nodes of the representative ATM network, (3) cumulative cells dropped at the separate buffers for each of the nodes of the ATM network, and (4) average number of cells dropped at the buffers throughout the entire network.

Figure 7.15 plots the throughput, expressed as a percentage, of the cells successfully transported across the switch fabric, for every node of the network. The average throughput over all nodes of the network is at 774,139 cells or 57.699%, implying that nearly half of the cells encountered by the respective switch fabric are lost due to buffer overflow. Clearly, poor QoS guarantees are inevitable. However, the lowest throughput of 35% at node 10 implies severe performance degradation and is the direct consequence of the traditional output buffer architecture, in concert with the choice of buffer size and the input traffic pattern. The wide range of variation of the throughput, namely 75% - 35% = 40%,

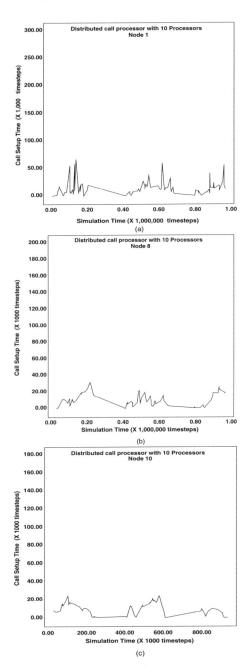

Figure 7.14 Call setup time as a function of simulation time, in timesteps, for (a) source node 1 to destination node 2, (b) source node 8 to destination node 9, and (c) source node 10 to destination node 12, for the 15-node representative ATM network under significant stress.

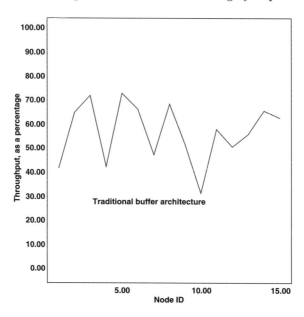

Figure 7.15 Throughput, i.e., cells successfully transported across the switch fabric, of every node.

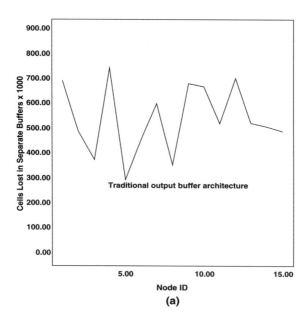

Figure 7.16 Cumulative number of cells dropped in the output buffers, as a function of the node identifier.

reflects the severity with which the statically determined fixed-size output buffers of nodes are overwhelmed, from time to time, with excessive outgoing cells.

As further corroboration of the network behavior, Figure 7.16 plots the cumulative number of cells dropped in the output buffers as a function of the node identifier. The cell loss is very high, and the general behavior corresponds to that for throughput, as shown in Figure 7.15. The cumulative number of cells dropped per node is 567,523 with the percentage value being 42.30%.

Clearly, the results in Figures 7.15 and 7.16 could further worsen should a perpetrator launch a coordinated attack from multiple geographically dispersed points, targeting one or more specific links in the ATM network and forcing an excessive number of call requests through the link(s). Even when the cumulative average bandwidth requests of these calls does not exceed the link's capacity, the high bandwidth utilization coupled with the bursty traffic may induce undue stress on the output buffer, causing significant cell loss.

7.3 Problems and Exercises

1. Determine, starting from the fundamentals, one additional source of complex vulnerability in ATM networks.

2. In IP networks, each packet must be self-contained in that any IP router node of the network must be able to process it completely on its own. That is, the network must be able to identify each packet, in a globally unique manner, through a simple or complex function of the source and destination IP addresses and other parameters. In contrast, in an ATM network, given that the transport of a cell is preceded by the call setup process, a set of piecewise locally unique identifiers replaces the need for a globally unique identifier. Discuss, from the security perspective, the relative advantages and disadvantages of these two contrasting approaches.

8
Future Issues in Information Systems Security

Given the incessant proliferation of information system networks into everyday life and the fact that by definition, a network's resources are shared among its users, network security will continue to play a dominant role. Conceivably, the future will witness networks that encompass very large geographical distances, support enormous increases in the numbers of users, nodes, and links, and offer highly sophisticated services, all of which will impose a greater demand on security. In addition to cryptography, other traditional areas of study, and those already elaborated in this book, the author proposes the pursuit of research and related activities in the future in the following important areas:

1. **Performance stability of networks under external perturbations:** Increasingly, experts in information systems are recognizing the vital role of system availability or stability, a concept that has long been recognized in the electric power community. If an enemy succeeds in degrading the control network sufficiently, none of the sophisticated cryptographic schemes are effective, since no messages will get across the networks. The nation's well-being is at risk if an enemy is capable of rendering the communications and control network, power grid [164], or banking network unavailable, without engaging the traditional defense forces or firing a single shot.

   Recently, Lee and Ghosh [165][166] addressed, from a practical perspective, the fundamental property of stability in asynchronous distributed decision-making (ADDM) systems, which subsumes information systems and networks. They report a thorough survey of the research efforts in stability in the relevant scientific disciplines including control systems engineering and introduce an empirical approach to relate stability to network performance, utilizing Lyapunov's work on stability of motion.

2. **Self-healing algorithms under link and node failures:** Military and civilian networks must offer high reliability despite inevitable partial failures that will develop from exposure to stress in the real world and battlefield situations. The problem is especially serious and challenging in high-speed networks, given the very high bandwidths, the promise of very low cell loss, and where users are offered statistical performance guarantees. The high operational speeds of networks render any manual restoration techniques unacceptably slow. Self-healing network control refers to the network's efforts of correcting or compensating so as to restore itself to its original integrity automatically, quickly, and autonomously.

3. **Design of innovative hardware architectures to address the increased computational demand from sophisticated, computationally intensive security functions:** Security measures, especially sophisticated procedures to prevent attacks, generally require significant computation. Given that the future will demand complex functionality, while network speed is likely to continue to grow, it is critical to develop novel hardware architectures for both the switching fabric and the processing engines.

4. **Distributed visualization of key security and performance parameters in operational networks:** Analysis of the literature reveals that the issue of visual representation of complex distributed algorithms, including networks, is critical. The reasons are as follows. First, it greatly facilitates the detection of design errors in data representation, data propagation, and resource allocation algorithms. Second, it assists in obtaining important performance parameters for the system and its underlying algorithms. Third, it helps develop a conceptual understanding of the algorithm's behavior for a given input stimulus and confirm the functional correctness of the implementation. In general, human thought processes are sequential, and it is relatively easy to conceptualize the operation of a uniprocessor executing a sequence of statements one at a time. However, where a large number of processors execute concurrently, asynchronously, and cooperatively and exchange information asynchronously, as in a real-world network, to solve a problem, the task of conceptualizing the algorithm's behavior becomes difficult. Visual displays can help immensely. Given the significant computational burden associated with visualization of a large network and the inherent geographic dispersion of its elements, distributed visualization [167] is important. In addition, it is proposed that the quality [168] of the decisions computed by each of the nodes of the geographically dispersed network be computed and displayed.

5. **Localizing sources of attacks, quickly:** In telephony, ISDN, and IP networks, to cause appreciable damage, an attack must persist for sufficient duration, often betraying its location in the process. In contrast, in ATM and future high-speed networks, the norm of high user bandwidths implies that extensive damage may result even when an attack persists only for a relatively short duration. Under these circumstances, coordinated attacks

that persist for extremely short durations, in the range of microseconds, or precision attacks that are timed to take effect immediately following specific network events, will resist detection. Experimental results in Chapters 6 and 7 appear to indicate that distributed detection techniques may hold promise in quickly recognizing the onset of such attacks and localizing their sources. Clearly, a systematic effort to study such techniques to permit fast yet accurate detection is warranted.

6. **Distributed architecture for intrusion detection:** The spirit of engineering lends us the belief that we should be able to construct an NIS system as a fortress that by design is impregnable to external attacks. Nonetheless, the possibility that NIS systems may be successfully penetrated from the outside or that one or more of the internal components may go haywire, either on their own or following an infection from an intrusion attack, must be considered during NIS systems design. Sandia National Laboratories [169] in the US has designed an intrusion-detection scheme for small-scale ATM networks where PNNI and UNI sensors are deployed at the edges of a network to enable an assessment engine to detect attacks by matching anomalous network behavior to standard templates gathered from known attacks and then deploying effective countermeasures via the response agents. While the approach realizes the sensors, assessment engines, and response agents in software, and is adequate for small-scale ATM networks, it also motivates the development of a practical, scalable intrusion-detection architecture for large-scale ATM networks. The causes and origins of the key challenges are enumerated as follows. First, intrusion attacks directed against the PNNI protocol, including the call admission control (CAC) and operation and management (OAM) functions, generally occur at speeds of milliseconds to seconds, which for small-scale ATM networks may be adequately addressed in software. Second, for modest to large-scale ATM networks, organized in the form of a collection of peer groups, the computation underlying the cumulative decision-making corresponding to the increased number of ATM nodes, may render a straightforward software solution impossible. Third, given that the processes in the ATM switch fabric operate at microsecond speeds or faster, intrusion attacks directed at ATM cell transport may defy a pure software solution. While a hardware-dominated approach may constitute a logical response to the latter two issues, there is an added concern. As ATM switching speeds continue to increase in the future, even a pure hardware solution may not be adequate. One must pursue a new approach, a novel decentralized architecture for intrusion detection in ATM and future high-speed networks.

7. **Lessons from nature's human immune system design:** The design of the human immune system represents nature's balanced effort to secure the complex human body from the adverse influence of foreign microbes while ensuring the continued existence of the microbes that cannot survive without complex hosts. Nature's efforts are exceptionally meticulous and span mil-

lions of years, implying that not only is the human immune system design a gold mine of intricate security principles but that it constitutes an incredible source of inspiration for developing a unique security architecture. Clearly, analysis of this unprecedented system, from the perspective of networked information systems security, will yield valuable insights.

8. **The practicality of network redesign for security:** There is undeniably a widespread belief among many NIT policymakers that the current IP networking technology has proliferated into society so deeply that despite its fundamental difficulty with security, any attempt to replace it with a superior design is economically futile. There are three key arguments against this thinking. First, history teaches us that in the long run, it is penny wise and pound foolish to reject new ideas and innovations and remain stationary. Consider the following instructive anecdote. After the American Revolution, the US Congress debated whether to introduce a new currency. Given the lack of funds in the treasury and the high cost of introducing and minting a new currency, the consensus was to adopt either the French or Italian currency. Thomas Jefferson strongly disagreed and asked Congress to accept the financial burden and introduce a new currency. He argued that the US could not aspire to become a great nation without its own currency. Fortunately, Congress shared Jefferson's vision, and while the US Mint had very humble beginnings, today, the United States without the US dollar is almost unimaginable. The author humbly envisions that in the not-so-distant future, a world without NIT systems will similarly appear inconceivable. Therefore, starting today, our thinking should be oriented toward designing the right kinds of NIT systems. Second, the danger from succumbing to complacency and allowing technology to stagnate is very real. Third, every network consists of two principal elements: the traffic-carrying links in the form of copper cables and optical fibers, commonly referred to as the infrastructure, and the switches or nodes that are housed in the closets of office buildings or switching stations. In general, migrating to a new networking technology requires one to replace only the switches; the underlying infrastructure remains functional. While an IP switch may be priced at around \$10,000, a comparable ATM switch may cost \$12,000. Even if the switch corresponding to a new networking principle were to cost as much as \$14,000, the cost of replacing the switches is insignificant when compared to the cost of the infrastructure and the benefit to society.

9. **Reliable and complete scientific path from conception to a reliable prototype system design:** Although security permeates numerous fields including banking, transportation, medicine, communications, and remote control, the security needs in each of these fields are probably unique. To realize a genuine revolution in secure NIT systems design, it is imperative that a scientific yet practical approach be developed that enables a user to start with an innovative solution to a security problem, validate it through computer modeling and simulation, and culminate in a hardware-

and software-based prototype system. In the event that the final design differs from the original specifications, there is a logical and systematic path that the user may follow to determine the cause of the discrepancy and correct it.

10. **A philosophical, sociological, judicial, and scientific framework to serve as a basis to help define, identify, and prosecute networked information systems crime:** Given the proliferation and importance of NIT systems and the current state of NIT systems security, the potential for harm to individuals, society, and nations is immense. For the information age to gain widespread acceptance in the world, a key requirement is to strike the right balance between the security needs of the individual person and those of the society as a whole. A principal vehicle to help realize this balance may consist of a set of precisely written laws, which must be developed taking into consideration philosophical, sociological, scientific, and, of course, judicial issues. At the present time, i.e., in 2001, US laws that pertain to network-related crime are, at best, in their infancy. In the absence of a holistic and logical basis, the design and interpretation of today's laws are linked solely to the monetary consequences of the crime, computable based only on today's understanding of the impact of the crime in terms of property and financial losses. The future monetary consequences of the crime, activities that weave the seeds of future harm into networks but do not inflict any monetary loss at the present, and other irreversible nonmonetary damages arising from the act of the crime, do not even factor into the current laws.

11. **Security in generalized networks:** Motivated by need and driven by demand, the discipline of networking is slowly but surely heading in the direction of "generalized networks," where the underlying architecture is algorithmic, precise, and assumes that every network element is mobile. Although cell phones constitute a manifestation of mobile wireless communication, the notion of terrestrial stationary network nodes as an indispensable infrastructure is deeply embedded within us. However, as soon as one visualizes the concept of networking in outer space, the notion of generalized networks hits home. In outer space, everything is in relative motion, the moon, planets, space stations, satellites, and spacecraft, implying a highly dynamic topology. Not even the familiar Earth is stationary anymore. Generalized networks are more prone to vulnerabilities than fixed networks, stemming from the attribute of mobility, and this warrants intense investigation. The wireless characteristic, by itself, produces only modest vulnerability. For although wireless transmission can sometimes be broadcast, spilling the communication into every corner of space, microwave, laser, and other high-frequency electromagnetic radiation may be focused into extremely sharp directional beams, very similar to communication along a wire.

12. **Maintaining a security edge through the practice of creativity:** To date, the history of research in security has been primarily reactive, i.e., it has been driven by the need to "catch up" with the breaches forced upon us

by smart perpetrators. Given the growing complexity of NIT systems, their enormous value, and the increasing sophistication of the perpetrators, the game of trying to play "catch up" is likely to become more and more difficult for organizations with significant stake in the integrity of NIT systems. The discovery of new and imaginative approaches that transcend current thinking by a quantum leap through the pursuit of creative exercises offers a promising solution [170][171]. Although the exact definition of creativity is elusive, approaches that can potentially trigger creativity, at will, in ordinary NIT personnel, warrant dedicated investigation.

13. **Key to long-term success in networked information systems security:** Given the logical observation that the discipline of networked information systems security is here to stay, a key to a successful strategy lies in combining research with educational initiatives. While research promises to uncover new ideas in security, teaching them in a class setting offers the unique opportunity to obtain immediate and valuable feedback from graduate and undergraduate students and practitioners. Their critiques, in turn, may be used to refine the original research ideas and conduct new experiments, or may lead to radically new thinking, eventually constituting new knowledge. Along with a number of leading educators, the author believes that a synergistic integration of recent advances in key areas including networking, network security, asynchronous distributed algorithms for coordination and control, innovative performance metric design, behavior modeling and asynchronous distributed simulation, reconfigurable hardware design, computational intelligence, data structures and algorithms, and programming principles, culminating in a new "networked information systems engineering program," may serve to train a new breed of personnel who will become future thinkers and leaders in the NIT field. An experiment in this direction is currently in progress in the Electrical and Computer Engineering Department at Stevens Institute of Technology. In concert, an institute for cybersecurity is also being established with the vision of a unique long-term synergistic integration of research, teaching, and training in networked information systems security. A key mission is to encompass as many sub-disciplines of knowledge as possible that contribute to holistic networked information systems security.

References

[1] Carl von Clausewitz. *Historical and Political Writings*. Edited and Translated by Peter Paret and Daniel Moran. Princeton University Press, Princeton, NJ, 1992.

[2] David Kahn. *The Codebreakers: The Story of Secret Writing*. Tun Huang Shu Chu Publishers, Taiwan, Reprint. Originally Published by Macmillian, New York, 1967, 1968.

[3] R.T. Marsh. Executive Summary, President's Commission on Critical Infrastructure Protection. Technical report, Washington, DC, October 13, 1997.

[4] J. Backhouse and G. Dhillon. Managing computer crime: A research outlook. *Computers and Security*, Vol. 14(7):645–651, 1995.

[5] History Channel. *Sworn to Secrecy*. Cable Television, April 18, 2000.

[6] History Channel. *Suicide Missions*. Cable Television, June 12, 2000.

[7] The Learning Channel. *Extreme Machines*. Cable Television, June 4, 2000.

[8] Seshasayi Pillalamarri and Sumit Ghosh. Fundamental Attributes of High-Speed Networks. Submitted to IEEE/ACM Transactions on Networking, December 2000.

[9] Sumit Ghosh. Private communications with Dan Olmos, Boeing Corporation, Sunnyvale, CA., February 2000.

[10] S. Lubkin. Asynchronous Signals in Digital Computers. *Mathematical Tables and Other Aids in Computation*, Vol. 6(40):238–241, October 1952.

[11] T.J. Chaney and C.E. Molnar. Anomalous behavior of synchronizer and arbiter circuits. *IEEE Transactions on Computers*, Vol. C-22(4):421–422, 1972.

[12] M.J. Stucki and J.R. Cox. Synchronization Strategies. In *Proceedings of the Caltech Conference on VLSI*, pages 375–386, January 1979.

[13] W. Stallings. *Cryptography and Internetwork Security – Principles and Practice*. Prentice Hall, NJ, 1999.

[14] C.P. Pfleeger. *Security in Computing.* Prentice Hall, NJ, 2nd Edition, 1997.

[15] Gregory White, Eric A. Fisch, and Udo W. Pooch. *Computer Systems and Network Security.* CRC Press LLC, Boca Raton, Florida, 1996.

[16] Patrica Edfors. Speaker on 21 March 1996, Network Rating Model Conference. Technical report, US Department of Justice, Williamsburg, VA, 1996.

[17] T.W. Madron. *Network Security in the '90s – Issues and Solutions for Managers.* John Wiley & Sons, Inc., New York, 1992.

[18] F. Simonds. *Network Security.* McGraw-Hill, New York, 1996.

[19] J. Hitchings. Deficiencies of the Traditional Approach to Information Security and the Requirements for a New Methodology. *Computers and Security,* Vol. 14(5):377–383, 1995.

[20] C.C. Baggett. Keynote address at the Network Rating Model, First Public Workshop. Technical report, US National Security Agency, Williamsburg, VA, March 20–22, 1996.

[21] Department of Defense. Department of Defense Trusted Computer System Evaluation Criteria, 5200.28-STD. Technical report, US Department of Defense, Washington, DC, 1985.

[22] National Computer Security Center. Trusted network interpretation of the trusted computer system evaluation criteria, NCSC-TG 005, ISBN 306-A-19. Technical report, US Department of Defense, Ft. George Mead, Maryland, July 31, 1987.

[23] W. Schwartau. *Information Warfare.* Thunder's Mouth Press, New York, 1996.

[24] Ira Winkler. *Corporate Espionage.* Prima Publishing, Rocklin, CA, 1997.

[25] G. White, E. Fisch, and U. Pooch. Government-Based Security Standards. *Information Systems Security,* pages 9–19, Fall 1997.

[26] R. Power. CSI Special Report on Information Warfare. *Computer Security Journal,* Vol. XI(2):63–73, 1995.

[27] M.D. Abrams and M.V. Joyce. Trusted System Concepts. *Computers and Security,* Vol. 14(1):45–56, 1995.

[28] D.M. Nessett. Layering central authentication on existing distributed system terminal services. In *Proceedings of the IEEE 1989 Computer Society Symposium on Security and Privacy,* pages 290–299, Oakland, CA., May 1–3, 1989.

[29] P. Lin and L. Lin. Security in Enterprise Networking: A Quick Tour. *IEEE Communications Magazine*, pages 56–61, January 1996.

[30] P. Janson and R. Molva. Security in Open Networks and Distributed Systems. *Computer Networks and ISDN Systems*, Vol. 22:323–346, January 1991.

[31] D.E. Geer. Electronic Commerce, Banking and You. *Computer Security Journal*, Vol. XI(2):55–62, 1995.

[32] H.H. Hosmer. Security is Fuzzy! Applying Fuzzy Logic to the Multipolicy Paradigm. *Computer Security Journal*, Vol. XI(2):35–45, 1995.

[33] S. Hill and M. Smith. Risk Management and Corporate Security. *Computers and Security*, Vol. 14(3):199–204, 1995.

[34] T. Chambers. Case study: A managerial perspective on an Internet security incident. *Computer Security Journal*, Vol. XI(1):17–23, 1995.

[35] H.B. Wolfe. Computer Security: For Fun and Profit. *Computers and Security*, Vol. 14(2):113–115, 1995.

[36] C. Oliver. Privacy, Anonymity, and Accountability. *Computers and Security*, Vol. 14:489–490, 1995.

[37] R.S. Vaccaro and G.E. Liepins. Detection of ANOMALOUS Computer Session Activity. In *Proceedings of the IEEE 1989 Computer Society Symposium on Security and Privacy*, pages 280–289, Oakland, CA., May 1–3, 1989.

[38] P. Helman and G. Liepins. Statistical Foundations of Audit Trail Analysis for the Detection of Computer Misuse. *IEEE Transactions on Software Engineering*, Vol. 19(9):886–901, 1993.

[39] T.F. Lunt and R. Jagannathan. A Prototype Real-Time Intrusion-Detection Expert System. In *Proceedings of the IEEE 1988 Computer Society Symposium on Security and Privacy*, pages 59–66, Oakland, CA., April 18–21, 1988.

[40] S. Kumar and E.H. Spafford. An application of pattern matching model in intrusion detection. Technical report 94-013, Department of Computer Sciences, Purdue University, Lafayette, Indiana, March 1994.

[41] B.C. Soh and T.S. Dillon. Setting optimal intrusion-detection thresholds. *Computers and Security*, Vol. 14(7):621–631, 1995.

[42] T.F. Lunt. A survey of intrusion detection techniques. *Computers and Security*, Vol. 12(4):405–418, 1995.

[43] S.A. Weerasooriya, M.A. El-Sharkawi, M. Damborg, and R.J. Marks II. Towards static-security assessment of a large-scale power system using neural networks. *IEE Proceedings, Part C, Generation, Transmission, and Distribution*, Vol. 139(1):64–70, January 1992.

[44] Mansour Esmaili, Reihaneh Safavi-Naini, and M. Bala Balachandran. AUTOGUARD: A continuous case-based intrusion detection system. *Australian Computer Science Communications*, Vol. 19(1):392–401, 1997.

[45] Andrew P. Kosoresow and Steven A. Hofmeyr. Intrusion detection via system call traces. *IEEE Software*, Vol. 14(5):35–42, September–October 1997.

[46] Shiuh-Pyng Shieh and Virgil D. Gligor. On a pattern-oriented model for intrusion detection. *IEEE Transactions on Knowledge and Data Engineering*, Vol. 9(4):661–667, July–August 1997.

[47] Nicholas Puketza, Mandy Chung, Ronald A. Olsson, and Biswanath Mukherjee. Software platform for testing intrusion detection systems. *IEEE Software*, Vol. 14(5):43–51, September-October 1997.

[48] G. Prem Kumar and P. Venkataram. Security management architecture for access control to network resources. *IEE Proceedings: Computers and Digital Techniques*, Vol. 144(6):362–370, November 1997.

[49] Diheng Qu, Brian M. Vetter, Feiyi Wang, Ravindra Narayan, S. Felix Wu, Y. Frank Jou, Fengmin Gong, and Chandru Sargor. Statistical anomaly detection for link-state routing protocols. In *International Conference on Network Protocols Proceedings of the 1998 International Conference on Network Protocols*, pages 62–70, Austin, TX, October 13–16, 1998.

[50] Jose Mauricio Jr. Bonifacio, Adriano M. Cansian, Andre C.P.L.F. de Carvalho, and Edson S. Moreira. Neural networks applied in intrusion detection systems. In *IEEE International Conference on Neural Networks–Conference Proceedings IEEE World Congress on Computational Intelligence Proceedings of the 1998 IEEE International Joint Conference on Neural Networks. Part 1 (of 3)*, pages 205–210, Anchorage, Alaska, May 4–9, 1998.

[51] Midori Asaka. Information gathering with mobile agents for an intrusion detection system. *Systems and Computers in Japan*, Vol. 30(2):31–37, February 1999.

[52] David Newman, Tadesse Giorgis, and Farhad Yavari-Issalou. Intrusion detection systems: Suspicious finds. *Data Communications*, Vol. 27(11):8, August 1998.

[53] Teresa Lunt. Panelist. Technical report, National Information Systems Security Conference, Baltimore, MD, October 8, 1997.

[54] R. Billington and E. Khan. A Security Based Approach to Composite Power System Reliability Evaluation. *IEEE Transactions on Power Systems*, Vol. 7(1):65–71, February 1992.

[55] Computer Sciences Corporation. UCA and DIAS Information Security Analysis. Technical Report EPRI Technical report TR-103773, Electric Power Research Institute, Palo Alto, CA, August 1994.

[56] S.A. Klein and J.N. Menendez. Information security considerations in open systems architecture. *IEEE Transactions on Power Systems*, Vol. 8(1):224–229, February 1993.

[57] J.A. Pecas Lopes, F.P. Maciel Barbosa, J.P. Marques de Sa, and J.M.G. Sa da Costa. A new approach for transient security assessment and enhancement by pattern recognition. In *Proceedings of the Second European Workshop on Fault Diagnostics, Reliability, and Related Knowledge Based Approaches*, pages 189–215, April 6–8, 1987.

[58] S.K. Fitzpatrick and P.J. Hargaden. Multimedia communications in a tactical environment. In *Proceedings of the IEEE MILCOM*, volume Vol. 1, pages 242–246, 1994.

[59] W. Morris, ed. *The American Heritage Dictionary of the English Language.* Houghton Mifflin Company, Boston, MA., 1981.

[60] *Webster's Ninth New Collegiate Dictionary.* Merriam Webster, Springfield, MA., 1985.

[61] Sumit Ghosh. *Hardware Description Languages: Concepts and Principles.* IEEE Press, 2000.

[62] L. Guidoux. Intelligent Solutions for Data Communications Networks. *Telecommunications, International Edition*, Vol. 29(6):25–29, June 6 1995.

[63] R.M. Needham and M.D. Schroeder. Using encryption for authentication in large networks of computers. *Communications of the ACM*, Vol. 21(12), December 1978.

[64] W. Diffie, M. Wiener, and P. Van Orshoot. *Authentication and authenticated key exchanges, designs, codes, and cryptography.* Kluwer Academic Press, Boston, MA, 1992.

[65] W. Stallings. *Network and Internetwork Security – Principles and Practice.* Prentice Hall, Englewood Cliffs, New Jersey, 1992.

[66] J. Kohl and B. Neuman. *Kerberos Network Authentication Service (V5).* DDN Network Information Center, September 1993.

[67] V.L. Voydok and S.T. Kent. Security mechanisms in high-level network protocols. *ACM Computing Surveys*, pages 135–171, June 1983.

[68] K. Thompson. On trusting trust. *Unix Review*, Vol. 7(11):71–74, 1984.

[69] J.D. Saltzer and M.D. Schroeder. The protection of information in computer systems. *Proceedings of the IEEE*, Vol. 63(9):1278–1308, March 1975.

[70] F.T. Grampp and R.H. Morris. Unix operating system security. *AT&T Bell Laboratories Technical Journal*, Vol. 63(8, Part 2):1649–1672, October 1984.

[71] Cylink Corporation. Cylink and GTE Announce First Successful Demonstration of Secure Video Teleconferencing! Infoguard 100 Proves Commercial Viability of Secure ATM Networks. Technical report, June 21, 1996.

[72] G.A. Spanos and T.B. Maples. Security for Real Time MPEG Compressed Video in Distributed Multimedia Applications. In *Proceedings of the IEEE 15th Annual International Phoenix Conference on Computers and Communications*, pages 72–78, Scottsdale, AZ, March 27–29, 1996.

[73] S-C. Chuang. A flexible and secure multicast architecture for ATM networks. In *IEEE Globecom*, pages 701–707, Singapore, November 14–16, 1995.

[74] C.A. Wilcox. ATDNet research at the National Security Agency. *IEEE Network*, pages 42–47, July/Aug 1996.

[75] R.H. Deng, L. Gong, and A.A. Lazar. Securing Data Transfer in Asynchronous Transfer Mode Networks. In *IEEE Globecom*, pages 1198–1202, Singapore, November 14–16, 1995.

[76] D. Stevenson, N. Hillery, and G. Byrd. Secure communications in ATM Networks. *Communications of the ACM*, Vol. 38(2):45–52, February 1995.

[77] M. Peyravian and T.D. Tarman. Asynchronous Transfer Mode Security. *IEEE Network*, Vol. 11(3):34–40, May/June 1996.

[78] P.W. Dowd and J.T. McHenry. Network security: It's time to take it seriously. *IEEE Computer*, Vol. 31(9):24–28, September 1998.

[79] B. Schneier. Cryptographic design vulnerabilities. *IEEE Computer*, Vol. 31(9):29–33, September 1998.

[80] A.D. Rubin and D.A. Geer. A survey of web security. *IEEE Computer*, Vol. 31(9):34–41, September 1998.

[81] R. Oppliger. Security at the Internet layer. *IEEE Computer*, Vol. 31(9):43–47, September 1998.

[82] T.D. Tarman, R.L. Hutchinson, L.G. Pierson, P.E. Scholander, and E.L. Witzke. Algorithm-agile encryption in ATM networks. *IEEE Computer*, Vol. 31(9):57–64, September 1998.

[83] K. Sato, S. Ohta, and I. Tokizawa. Broad-Band ATM Network Architecture Based on Virtual Paths. *IEEE Transactions on Communications*, Vol. 38(8):1212–1222, August 1990.

[84] Anna Hac and Hasan B. Mutlu. Synchronous Optical Network and Broadband ISDN Protocols. *IEEE Computer*, Vol. 22(11):26–34, November 1989.

[85] Arthur Chai and Sumit Ghosh. Modeling and Distributed Simulation of Broadband-ISD Network on a Network of Sun Workstations Configured as a Loosely-Coupled Parallel Processor System. *IEEE Computer*, Vol. 26(9):37–51, September 1993.

[86] ATM Forum Technical Committee. Private Network–Network Interface Specification Version 1.0 (PNNI 1.0). www.atmforum.com/atmforum/specs/approved.html, ATM Forum, March 1996.

[87] E.W. Dijkstra. A Note on Two Problems in Connexion with Graphs. *Numerische Mathematik*, Vol. 1:269–271, 1959.

[88] L. Runge. Security and the Transition to Client/Server Computing. *Information Systems Security*, pages 49–57, Spring 1996.

[89] R. Hale. End-User Computing Security Guidelines. *Information Systems Security*, pages 49–64, Winter 1996.

[90] T. Stacey. The Information Security Program Maturity Grid. *Information Systems Security*, pages 22–33, Summer 1996.

[91] D.P. Bertsekas and R.G. Gallager. *Data Communication Networks*. Prentice Hall, Englewood Cliffs, NJ, 1987.

[92] P. Ferguson and G. Huston. *Quality of Service*. John Wiley and Sons, Inc., New York, 1998.

[93] Sumit Ghosh and Tony Lee. *Principles of Modeling and Asynchronous Distributed Simulation of Complex Systems*. IEEE Press, NJ, 2000.

[94] Sumit Ghosh and Pete Robinson. A Framework for Investigating Security Attacks in ATM Networks. In *Proceedings of the IEEE MILCOM'99 Conference*, pages 724–728, Atlantic City Convention Center, NJ, October 31–November 3, 1999.

[95] L. Kleinrock. The Latency/Bandwidth Tradeoff in Gigabit Networks. *IEEE Communications Magazine*, April 1992.

[96] N. Dorsen and S. Gillers, eds. *Government Secrecy in America: None of Your Business*. Viking Press, New York, 1973.

[97] F. Horton and D. Marchand, eds. *Information Management in Public Administration*. Information Resources Press, Arlington, VA, 1982.

[98] B. Mitchell and T. Donyo. Utilization of the U.S. Telephone Network. Technical Report HE8815.M58, Rand Corporation Study in cooperation with the European-American Center for Policy Analysis, 1994.

[99] J. Conover. ATM Backbone Switches: How Strong is Your Backbone? We Scrutinize 5 Switches. *Network Computing*, Vol. 8(21):78–89, November, 15 1997.

[100] AT&T. AT&T WorldNet Managed Internet Service Internet site: www.att.com/worldnet/wmis/misb.html. Technical report, April 1998.

[101] C.D. Chaffee. *The Rewiring of America The Fiber Optics Revolution*. Academic Press, New York, 1988.

[102] G.H. Johannessen. Signaling Systems: An International Concern. *Bell Laboratories Record*, Vol. 48(1):13–18, January 1970.

[103] C. Breen and C.A. Dahlbom. Signaling Systems for Control of Telephone Switching. *Bell System Technical Journal*, Vol. 39(6):1381–1444, November 1960.

[104] The Technical Journal. *http://mbay.net/~ mpoirier/bstj.html*, pages 1–3, August 2000.

[105] Qutaiba Razouqi, Seong-Soon Joo, and Sumit Ghosh. Performance Analysis of Fuzzy Thresholding-Based Buffer Management for a Large-Scale Cell-Switching Network. *IEEE Transactions on Fuzzy Systems*, Vol. 8(4):425–441, August 2000.

[106] V. Catania, G. Ficili, S. Palazzo, and D. Panno. Using fuzzy logic in ATM source traffic control: Lessons and perspectives. *IEEE Communications Magazine*, pages 70–81, November 1996.

[107] ATM Forum technical committee. Traffic management specification version 4.0. Technical Report 95-0013R9, ATM Forum, 1995.

[108] Fabrice Guillemin, Catherine Rosenberg, and Josee Mignault. On Characterizing an ATM Source via the Sustainable Cell Rate Traffic Descriptor. In *Proceedings of the IEEE Infocom'95, Session 96.2.1*, pages 1129–1134, 1995.

[109] J.W. Roberts. Traffic control in the B-ISDN. *Computer Networks and ISDN Systems*, Vol. 25:1055–1064, 1993.

[110] P. Pancha and M. El Zarki. Leaky bucket access control for VBR MPEG video. *Proceedings of IEEE INFOCOM '95*, April 1995.

[111] E. Rathgeb. Modeling and performance comparison of policing mechanisms for ATM networks. *IEEE Journal on Selected Areas in Communications*, Vol. 9(3):325–334, April 1991.

[112] D. Hong and J.J. Suda Te. Survey of techniques for prevention and control of congestion in an ATM network. *IEEE International Conference on Communications*, pages 204–210, April 1991.

[113] V. Catania, G. Ficili, S. Palazzo, and D. Panno. A Comparative Analysis of Fuzzy Versus Conventional Policing Mechanisms for ATM Networks. *IEEE/ACM Transactions on Networking*, Vol. 4(3):449–459, June 1996.

[114] J. Kurose. Open issues and challenges in providing quality of service guarantees in high speed networks. *Computer Communications Review*, pages 6–15, January 1993.

[115] D. Ferrari and D.C. Verma. A scheme for real time channel establishment in wide-area networks. *IEEE Journal on Selected Areas in Communications*, Vol. 8(3):368–379, April 1990.

[116] L. Zhang. Virtual clock: A new traffic control algorithm for packet switched networks. *ACM Transactions on Computer Systems*, Vol. 9(2):101–124, May 1991.

[117] D. Ferrari, A. Gupta, M. Moran, and B. Wolfinger. A continuous media communication service and its implementation. *Proceedings of GLOBE-COM '92*, December 1992.

[118] D.D. Clark, S. Shenker, and L. Zhang. Supporting real time applications in an integrated services packet network: Architecture and mechanism. *Proc. ACM SIGCOMM '92*, 1992.

[119] A.A. Lazar and G. Pacifici. Control of resources in broadband networks with quality of service guarantees. *IEEE Communications Magazine*, pages 66–73, October 1991.

[120] M.J. Hyman, A.A. Lazar, and G. Pacifici. Real time scheduling with quality of service constraints. *IEEE Journal on Selected Areas in Communications*, Vol. 9(7):1052–1063, September 1991.

[121] J.M. Peha. The priority token bank: Integrated scheduling and admission control for an integrated-service network. *IEEE International Conference on Communications*, pages 345–351, May 1993.

[122] H. Kreoner, M. Eberspeacher, T.H. Theimer, P.J. Keuhn, and U. Briem. Approximate analysis of the end to end delay in ATM networks. *Proceedings of IEEE INFOCOM '92*, 1992.

[123] S.J. Golestani. Congestion-free communication in high speed packet networks. *IEEE Transactions on Communications*, Vol. 39(12):1802–1812, December 1991.

[124] F. Gong and G. Parulkar. A two level flow control scheme for high speed networks. *Journal of High Speed Networks*, Vol. 3:261–284, 1994.

[125] C.M. Aras, J.F. Kurose, D.S. Reeves, and H. Schulzrinne. Real time communication in packet switched networks. *Proceedings of the IEEE*, Vol. 82(1):122–138, January 1994.

[126] D. Lutas, N. Linge, N. Robinson, E. Ball, J. Ashworth, and D. Gibbs. Multiprocessor system for interconnection of Ethernet and FDDI networks using ATM via satellite. *IEEE Proc.-Comput. Digit. Tech.*, Vol. 143(1):69–78, January 1996.

[127] Cristina Aurrecoechea, Andrew T. Campbell, and Linda Hauw. A Survey of QoS Architectures. In *Proceedings of the 4th IFIP Workshop on Quality of Service*, March 1996.

[128] Jean-Yves Le Boudec, Gustavo de Veciana, and Jean Walrand. QoS in ATM: Theory and Practice. In *Proceedings of Infocom'98*, 1998.

[129] C.S. Chang and J.A. Thomas. Effective bandwidth in high speed digital networks. *IEEE Journal of Selected Areas in Communication*, Vol. 13:1091–1100, August 1995.

[130] G.L. Choudhury, D.M. Lucanton, and W. Whitt. Squeezing the most out of ATM. *IEEE Transactions on Communication*, Vol. 44(2), February 1996.

[131] C. Tryfonas, A. Varma, and S. Varma. Efficient Algorithms for Computation of the Burstiness Curve of Video Sources. In *Proceedings of the International Teletraffic Congress (ITC)'99*, June 1999.

[132] T.V. Lakshman, P.P. Mishra, and K.K. Ramakrishnan. Transporting Compressed Video over ATM Networks with Explicit Rate Feedback Control. In *Proceedings of Infocom'97*, April 1997.

[133] R.P. Singh, S.H. Lee, and S.K. Kim. Jitter and Clock Recovery for Periodic Traffic in Broadband Packet Networks. *IEEE Transactions on Communication*, Vol. 42(5), May 1994.

[134] P.V. Rangan, S.S. Kumar, and S. Rajan. Continuity and Synchronization. *IEEE Journal of Selected Areas in Communication*, Vol. 14(1), January 1996.

[135] V.S. Frost and B. Melamed. Traffic modeling for telecommunications networks. *IEEE Communications Magazine*, pages 70–81, March 1994.

[136] CCITT Study Group XVIII. Traffic Parameters and Descriptors—Temporary Document 43. Technical report, CCITT, December 1990.

[137] H. Heffes and D.M. Lucantoni. A Markov modulated characterization of packetized voice and data traffic and related statistical multiplexer performance. *IEEE Journal on Selected Areas in Communications*, Vol. 4(6):856–868, September 1986.

[138] B. Maglaris, D. Anastassiou, P. Sen, G. Karlsson, and J.D. Robbins. Performance models of statistical multiplexing in packet video communications. *IEEE Transactions on Communications*, Vol. 36(7):834–844, July 1988.

[139] R. Gruenenfeld, J.P. Cosmas, S. Manthorpe, and A. Odinmaokafor. Characterization of video codecs as autoregressive moving average processes and related queueing system performance. *IEEE Journal on Selected Areas in Communications*, Vol. 9(3):284–293, April 1991.

[140] R.M. Rodriguez-Dagnino, M.R.K. Khansari, and A. Leon-Garcia. Prediction of bit rate sequences of encoded video signals. *IEEE Journal on Selected Areas in Communications*, Vol. 9(3):294–304, April 1991.

[141] S. Manthorpe, J. Schormans, J. Pitis, and E. Schaarf. A simulation study of buffer occupancy in the ATM access network: Are renewal assumptions justified? *International Teletraffic conference*, Vol. 13:801–805, 1991.

[142] R.C.F. Tucker. Accurate method for analysis of a packet speech multiplexer with delay. *IEEE Transactions on Communications*, Vol. 36:479–483, 1988.

[143] W.E. Leland, M.S. Taqqu, W. Willinger, and D.V. Wilson. On the Self-Similar Nature of Ethernet Traffic (Extended Version). *IEEE/ACM Transactions on Networking*, Vol. 2(1), February 1994.

[144] N.T. Plotkin and C. Roche. The Entropy of Cell Streams as a Traffic Descriptor in ATM Networks. In *Proceedings of the IFIP Performance of Communication Networks*, October 1995.

[145] E.E. Knightly and H. Zhang. D-BIND: An Accurate Traffic Model for Provisioning QoS Guarantees to VBR Traffic. *IEEE/ACM Transactions on Networking*, Vol. 5(No. 2):219–231, April 1997.

[146] G.D. Stamoulis, M.E. Anagnostou, and A.D. Georgantas. Traffic source models for ATM networks: a survey. *Computer Communications*, Vol. 17(6):428–438, June 1994.

[147] K. Sriram, R.S. McKinney, and M.H. Sherif. Voice packetization and compression of broadband ATM networks. *IEEE Journal on Selected Areas in Communications*, Vol. 9(3):294–304, April 1991.

[148] R. Jain and S. Routhier. Packet trains, measurements, and a new model for computer traffic. *IEEE Journal on Selected Areas in Communications*, Vol. 4(6):986–995, September 1986.

[149] Seong-Soon Joo and Sumit Ghosh. A New Metric Towards Comprehensive Performance Evaluation of ATM Networks. In *Proceedings of the IEEE ATM '98 Conference*, pages 368–372, George Mason University, VA, May 26–29, 1998.

[150] Seong-Soon Joo and Sumit Ghosh. Generalized End-to-End Performance Estimation of ATM Networks Using Combined Simulation Model and Neural Network. In *Proceedings of the IEEE Globecom'98 Conference, Session 116.7, Genetic Algorithms*, pages 116.7.1– 116.7.5, Sydney, Australia, November 8–12, 1998.

[151] Mesquite Software, Inc. User's Guide: CSIM18 Simulation Engine. Technical report, Mesquite Software, Inc., 1997.

[152] R.Y. Awdeh and H.T. Mouftah,. Survey of ATM Switch Architecture. *Computer Networks and ISDN Systems*, Vol. 27:1567–1613, July 1995.

[153] F. Kamoun, and L. Kleinrock. Analysis of shared finite storage in a computer network node environment under general traffic conditions. *IEEE Transactions on Communications*, Vol. 28(7):992–1003, July 1980.

[154] A.E. Eckberg and T.C. Hou. Effects of output buffer sharing on buffer requirements in an ATDM packet switch. *Proceedings of IEEE INFOCOM '88*, pages 459–566, March 1988.

[155] N. Endo, T. Kozaki, T. Ohuchi, H. Kuwahara, and S. Gohara. Shared Buffer Memory Switch for an ATM Exchange. *IEEE Transactions on Communications*, Vol. 41(1):237–245, January 1993.

[156] I. Iliadis. Performance of a packet switch with input and output queueing under unbalanced traffic. *Proceedings of IEEE INFOCOM '92*, Vol. 3:743–752, May 1992.

[157] A. Pattavina. Nonblocking architectures for ATM switching. *IEEE Communications Magazine*, (2), February 1993.

[158] M.A. Pashan, M.D. Soneru, and G.D. Martin. Technologies for broadband switching. *AT&T Technical Journal*, pages 39–47, November/December 1993.

[159] J. Causey and H. Kim. Comparison of Buffer Allocation Schemes in ATM Switches: Complete Sharing, Partial Sharing and Dedicated Allocation. *IEEE International Conference on Communications*, Vol. 2:1164–1168, 1994.

[160] MCI Systems Technical Documentation. *(www.mci.com)*, 1998.

[161] Fore Systems Technical Documentation. *(www.fore.com/atm-edu)*, 1996.

[162] Zbiquiew Dziong. *ATM Network Resource Management,*. *McGraw-Hill, Inc.*, 1997.

[163] C. Courcoubetis, G. Fouskas, and R. Weber. On the Performance of an Effective Bandwidth Formula. *Proceedings of ITC14*, pages 201–212, 1994.

[164] Reuters. Hackers hit computers running California's power grid. *Sacramento, CA*, pages 549–561, June 9, 2001.

[165] Tony Lee and Sumit Ghosh. On the Concept of "Stability" in Asynchronous Distributed Decision-Making Systems. In *Proceedings of the Fourth International Symposium on Autonomous Decentralized Systems, ISADS99*, pages 302–309, Tokyo, Japan, March 21–23, 1999.

[166] Tony Lee and Sumit Ghosh. On "Stability" in Asynchronous, Distributed, Decision-Making Systems. *IEEE Transactions on Systems, Man, and Cybernetics*, Vol. 30, Part B(4):549–561, August 2000.

[167] Tom Morrow and Sumit Ghosh. DIVIDE: Distributed Visual Display of the Execution of Asynchronous Distributed Algorithms on Loosely-Coupled Parallel Processors. *Computers and Graphics: An International Journal*, Vol. 18(6):849–859, December 1994.

[168] Tony Lee, Sumit Ghosh, and Anil Nerode. Asynchronous, Distributed, Decision-Making Systems with Semi-Autonomous Entities: A Mathematical Framework. *IEEE Transactions on Systems, Man, and Cybernetics*, Vol. 30, Part B(1):229–239, February 2000.

[169] Tom D. Tarman, Edward L. Witzke, Keith C. Bauer, Brian R. Kellogg, and William F. Young. Asynchronous Transfer Mode (ATM) Intrusion Detection. In *Proceedings of the IEEE MILCOM'01 Conference*, Atlantic City Convention Center, NJ, October 2001.

[170] Peter Heck and Sumit Ghosh. A Study of Synthetic Creativity through Behavior Modeling and Simulation of an Ant Colony. *IEEE Intelligent Systems*, Vol. 15(6):58–66, November/December 2000.

[171] Sumit Ghosh. Reflection as a Catalyst in Triggering Creativity in Science & Engineering: A Position Paper. *Submitted for Peer Review*, June 2001.

Index